MISUSED STATISTICS

POPULAR STATISTICS

D. B. Owen

Founding Editor

Additional Volumes in Preparation

MISUSED STATISTICS

SECOND EDITION, REVISED AND EXPANDED

Herbert F. Spirer
Columbia University, New York, New York, and
University of Connecticut, Stamford, Connecticut

Louise Spirer
Consultant
Stamford, Connecticut

A. J. Jaffe
Consultant
Leonia, New Jersey

MARCEL DEKKER, INC.   NEW YORK · BASEL · HONG KONG

ISBN: 0-8247-0211-5

This book is printed on acid-free paper.

Headquarters
Marcel Dekker, Inc.
270 Madison Avenue, New York, NY 10016
tel: 212-696-9000; fax: 212-685-4540

Eastern Hemisphere Distribution
Marcel Dekker AG
Hutgasse 4, Postfach 812, CH-4001 Basel, Switzerland
tel: 44-61-261-8482; fax: 44-61-261-8896

World Wide Web
http://www.dekker.com

The publisher offers discounts on this book when ordered in bulk quantities. For more information, write to Special Sales/Professional Marketing at the headquarters address above.

Current printing (last digit):
10 9 8 7 6 5 4 3 2 1

PRINTED IN THE UNITED STATES OF AMERICA

In Memoriam

To A. J. Jaffe, whose insight, knowledge, and wit made this book possible.

A. J. Jaffe died at age 85 on December 22, 1997,
just ten days after our last joint work session on the manuscript
for this second edition. This work is a summation of his life-long
dedication to searching out the truth in numbers.

Preface

Every day we see misuses of statistics that affect the outcomes of elections, change public policy, win arguments, get readers and viewers for newspapers and television programs, impress audiences, support prejudices, inflame hatreds, provoke fears, motivate governmental repressions, and sell products.

It doesn't have to be that way! Statistics—the science of using, summarizing, and analyzing numbers—*can be* and often *is* correctly used, and when it is, we are all better off. Good uses of statistics help to properly inform the public, puncture pretensions, contribute to the reduction of prejudices and hatreds, allay fears, deter aggression and repression, and help make products of high quality that can enrich lives.

We, the authors, are by nature measurers and questioners, which is why we are in this profession. As measurers, we contribute to the benefits of well-used statistics. As questioners, we work against the bad effects of misused statistics. Because we questioned, we started a series of articles on misuses of statistics that appeared for over ten years in the *New York Statistician*. These articles attracted great interest, including that of Maurits Dekker, who suggested we expand the series into a book. This first edition had a rewarding success in attracting readership and stimulating discussion.

Now, at the suggestion of Maurits Dekker's grandson, Russell Dekker, we have updated our examples, removed outdated ones, and added many new ones. Unfortunately, adding new misuses is all too easy, as there is a never-ending supply. More than 50% of this edition is new or revised material.

Our goal in this book is to help you to make sound judgments of statistical uses. We help you to know the right questions to ask by showing you dozens of examples of misuses of statistics, and providing some of the questions you can and must ask to see things as they are. As you will discover in reading this book, asking questions about statistics is a continuous and rewarding process. Every question you ask brings more information of value to you. We found this process to be as exciting as detective work and hope that you will too.

To help you, we have classified misuses into arbitrary categories and grouped them into chapters. You will also see that misuses often come in clusters, and sometimes so many misuses are in one analysis that we must call it a "megamisuse." You may think a classification other than ours is better. If you feel that way, good! This means that you are also becoming a questioner in this search after a special kind of truth.

Our book is not a textbook or report of investigations, which is where proper uses are and should be described. However, we have included some examples of the proper use of statistics to illustrate our point that good uses are available to everyone.

Don't let the presence of so many misuses in one book discourage you from using statistics or trusting statisticians! We all know that the proper use of statistics has been of great benefit in many areas. We hope that the number of such uses will be *increased* by your awareness of the misuses and good uses that we show here.

Have we ourselves misused statistics? We hope not, but, as you will see, it is hard to avoid an inadvertent misuse and we are as fallible as any other statistician.

We have prepared this book mainly for people whose statistical knowledge is less than that of the professional statistician, but it is meant to serve professionals as well. Because this book illustrates statistics as applied to real-world problems, it should be valuable as an adjunct to statistical textbooks, both graduate and undergraduate. And because we deal with real-world problems and events, it can also provide a valuable service for those who rely on statistical reports or follow national and international events in the media. We trust this text will lead our readers down the path of skepticism and questioning to understanding.

Herbert F. Spirer
Louise Spirer
A. J. Jaffe

Acknowledgments

We are indebted to many people—statisticians, scientists, friends, and others—who helped in so many ways with this second edition. Responding to our personal pleas, and to notices in *Chance* and *Amstat News*, many statisticians supplied examples that stimulated our thinking and challenged our approaches, and a number of these examples found their way into this book. Others were perceptive critics of our work and we hope that this book is more readable, understandable, and valuable for their assistance.

We list these supporters alphabetically, and can only hope that we have not missed anyone: David Banks, Peter Behuniak, Vincent J. Calabrese, Carl Castrogiovanni, Beth Osborne Daponte, Robyn M. Dawes, Marilynn Dueker, David J. Finney, Arnold Gordon, Tom Jabine, Sidney Jones, David Kerridge, Gary King, Yossi Levy, Robert Lewis, Daniel Levitin, Nick Longford, James Lynch, Matt MacIsaac, John R. Michael, Nathan Leon Pace, Mike Quinn, David Rios, Douglas E. Samuelson, Kelley Smith, Carl G. Sontheimer, Edward Spar, Paul A. Strassmann, Bernard W. Taylor III, Jim Turbitt, Harold Wainer, Chamont Wang, Angela Yorwerth, and Nancy Zager. We are grateful to Columbia University and the University of Connecticut in Stamford (and their excellent libraries) for their support. There are still others whose names you will see scattered throughout the text with their contributions.

We would also like to thank the many anonymous journalists, and government employees in the Bureau of the Census, the Federal Reserve Board, and the State of Connecticut and other agencies, who were remarkably patient and cooperative with our requests for information.

Contents

Contents

1
Introduction

You can fool all of the people some of the time, and some of the people all of the time, but you can't fool all of the people all of the time.
—Abraham Lincoln

Misuse of Statistics: Using numbers in such a manner that—either by intent, or through ignorance or carelessness—the conclusions are unjustified or incorrect.
—The Authors' Definition

NUMBERS COUNT

Numbers—all numbers—are important. Even a zero can be important. After all, the only difference between a poverty rate of 1% and a poverty rate of 10% is a single zero. And "statistics," the science of collecting, analyzing, and interpreting quantitative data, is how we can use data to determine accurately the placement of that zero.

Humans have used statistics for thousands of years, perhaps even before modern *Homo sapiens sapiens* appeared on the Earth. We know of censuses taken 3000 years ago, and numbers are recorded on the ancient tablets and stelae that date back to when written (or inscribed) symbols were first developed.

We have no doubt that prehistoric humans used quantitative thinking to decide whether the game the hunters brought home, or the stockpile of gathered fruits, vegetables, and nuts, was adequate to feed the group. Even if our ancestors did not use our sophisticated methods of counting, their very survival suggests that they estimated as we do. They must have compared the number of people with the amount of food available. At the kill of a mastodon, the members of a small group must have been able to see that there was more meat available than the group could eat. On the other hand, if the hunter brought back only one rabbit, he must have known that it was not enough for more than one or two people.

Today, we need more sophisticated methods for dealing with numbers because our concerns go beyond hunting and gathering. A restaurant owner, taking the place of our prehistoric hunter, must keep records of which dishes were the most popular, for a mistake of only a few percent could result in a refrigerator full of spoiled food.

A statistic of great importance to people in the United States is the number of people living in each state. The U.S. Constitution mandates that the members of the House of Representatives be apportioned among the states in accordance with each state's population. Thus, the U.S. Bureau of the Census counts the population every 10 years. At present, the bureau must use complex methods because of the large size of the population and the complexity of the task. Since these numbers determine the number of representatives to which a state is entitled, a mistake could result in improper representation and, of course, considerable political outcry. A current (as of 1997) dispute over whether to use statistical sampling along with simple enumeration in districts where accurate counts may be difficult to obtain in the year 2000 census, illustrates both the lack of knowledge of the accuracy of statistical sampling and the political fallout over who benefits from the count. We discuss this issue later, in Chapters 10 and 14.

Cruising on today's information highway, you cannot escape numbers, nor can you escape sophisticated statistical methods. Nor do most people want to. Imagine baseball without batting averages, number of bases stolen, runs scored, games won and lost, and pitching records. Or football without the number of yards gained, number of passes completed, lengths of passes, and so forth. People in sports who do not see themselves as "statisticians" continually make decisions based on statistics: the manager who has to decide who is the best available pinch hitter in a given situation, measures "best" in terms of batting averages under similar circumstances. Ranking the tennis players entered in the U.S. Open tournament from highest to lowest is the result of a complicated statistical formula.

If your interest is business and finance rather than sports, think of the importance of stock market indexes; the movement of commodity, stock, and bond prices; corporate profits or bond ratings; and other financial indicators. Decisions to buy or sell are made on the basis of the information conveyed by these statistics. Are you concerned with marketing? If so, you will need numbers and statistical methods to answer questions for making key decisions: How many potential customers are there for a particular product or in a particular geographic region? How rich or how poor are your potential and actual customers? What are the characteristics of your potential customers?

Today, investigators can use sophisticated technologies to analyze DNA to produce evidence for use in criminal investigations, and lawyers and judges may have to be trained to understand the implications of probability and statistics.

Numbers are the basis for political decisions on what actions to take and which policies to accept, reject, or formulate. Government officials constantly make decisions based on numbers. For example, the unemployment rate is watched by elected officials, administrators, and the public when deciding whether the government should take action to influence the economy: Shall we start a job development program to reduce unemployment in a particular region? Shall we pay more unemployment benefits for longer? The Consumer Price Index is a critical element in national budgetary expenditures and the well-being of many individuals. Should it be changed to reflect improvements in quality and changes in personal tastes? What effect will those changes have?

Skeptics may argue that political decisions are not made solely on the relevant numbers, but they should remember that many political decisions almost invariably take into account the number of voters with a given opinion. Elected officials are constantly checking the results of polls; indeed, poll results often seem to dominate the coverage of elections, and an elected official who votes against an issue that is supported by a large proportion of his constituents does so at his peril. For today's President, a resident pollster is a necessity.

Because there is so much numerical analysis going into the great majority of governmental decisions, a considerable number-crunching establishment (the U.S. Census Bureau, the U.S. Bureau of Labor Statistics, the General Accounting Office, and their equivalents in other countries) now exists. But this was not always the case. Below we show how a British nurse, Florence Nightingale, by inspired data collection and presentation, influenced her government to improve the conditions of those who fought in the Crimean War. Her work was truly remarkable.

The "Lady with a Lamp" Was a Numbers Cruncher

The statistical work of Florence Nightingale, founder of modern medical nursing, is a shining example of the *proper* use of statistics. Nightingale was appalled by the lack of medical and nursing care and the widespread presence of unsanitary conditions that she found in British army hospitals during the Crimean War. She collected and analyzed extensive data on these conditions. Her remarkable study and its successful presentation in Parliament as part of her advocacy led to medical and nursing programs which greatly reduced human suffering.

She collected data that showed decisively and quantitatively that more soldiers died from diseases caused by unsanitary conditions than from battle wounds. In January 1855, excluding men killed in action, the mortality rate "peaked at an annual rate of 1174 per 1000. . . . This means that if mortality had persisted for a full year at the rate that applied in January, and if the dead soldiers had not been replaced, disease alone would have wiped out the entire British army in the Crimea" (1, p. 133).

Her remarkably effective lobbying, supported by then novel graphic presentations of the mortality statistics, brought an immediate response from the British public and government. She was given government support to make sanitary reforms, which caused the death rate to drop dramatically. This she also documented superbly. Her statistical analyses were the key factor in motivating the modernization of the British armed forces medical programs and, subsequently, the introduction of a new standard of nursing and hospital care:

> Nightingale . . . strongly influenced the commission's work, [Royal Commission on the Health of the Army] both because some of its members were her friends . . . and because she provided it with much of its information. . . . She wrote and had privately printed an 800-page book [*Notes on Matters Affecting the Health, Efficiency and Hospital Administration of the British Army*] . . . which included a section of statistics accompanied by diagrams. Farr [a professional statistician] called it "the best [thing] that ever was written" either on statistical "Diagrams or on the Army." . . . Nightingale was a true pioneer in the graphical representation of statistics . . . much of Nightingale's work found its way into the statistical charts and diagrams [Farr] prepared for the final report. . . . Nightingale had the statistical section of the report printed as a pamphlet and widely distributed. . . . She even had a few copies of the diagrams framed for presentation to officials in the War Office and in the Army Medical Department [1, pp. 133, 136].

MISUSING STATISTICS: SOME FACTS OF LIFE

The purpose of our book is to show how to recognize misuses of statistics that lead to wrong decisions. What happens next is up to the reader. Below are the three points which form the heart of our philosophy.

1. *It is an unpardonable sin to misuse numbers for propaganda purposes.* The deliberate misuse of statistics to "prove" a point for political advantage often leads to policies which cause serious harm to peoples, organizations, and countries. This even applies to some of those organizations known as "nongovernmental organizations" (NGOs), formed for some public

good purpose, which sometimes publish statistics that are more useful to their advocacy positions than truthful.

    2. *Numbers need only be as correct as is necessary for the purpose at hand.* For example, during the period 1970–1990, uncounted and uncountable numbers of land mines were laid in Cambodia. Tens of thousands of Cambodians are known, through hospital records of their treatment, to have suffered limb amputations as a consequence. It is likely that many others either died or did not receive treatments. Is it essential to have an exact measure of the number of the amputees to know that action must be taken and to estimate an appropriate level of response? Ten child amputees can be enough to prove the point.

    3. *Most statistics are not exact and may not be exact enough for the purposes at hand.* But how inexact? Sometimes we need highly accurate statistics; if the statistics you have are exact enough, you can make a decision; otherwise, you can't. For example, in the apportionment of representatives to the states, the loss or gain of a congressman can occur for a difference of 1000 people in a state with a population in the millions. In relative terms, an error of 1000 out of 10,000,000 amounts to an error of one part in 10,000. This looks like an extremely small error, and it would be in most circumstances. But here it looms large because of its possible consequences. In other cases, such exactitude is less important. The *exact* number of refugees in a camp in the Democratic Republic of Congo may not be essential. An estimate is enough to compute how many sacks of rice are necesssary.

WHY KNOWLEDGE OF MISUSES IS IMPORTANT

Learning from Misuses

Be determined! We can all profit from the study of misuses and misinterpretations of statistics. The general public, elected and appointed governmental officials, business people, lawyers and judges, journalists, TV and radio commentators, medical personnel, dentists, and students can learn how to detect a misuse of statistics. Even professional statisticians and teachers can profit, for none of us knows all that there is to know.

    This book's goal for nonstatisticians is to show how to spot misuses in the media (including scholarly and professional journals) and in public statements. The key question for such readers is: *What (or whom) shall I believe?* This question is best answered when the questioner can evaluate statistical arguments.

    Our goal for professional statisticians (including teachers) and those who must interpret data and publish articles that include or are based upon statistics, is to avoid statistical misuse. Analysts and report writers often fool themselves, as well as the public, by misusing statistics through inadvertence, ignorance, or neglect. Even highly skilled professionals make errors, as will be seen when we discuss the remarkable case of Samuel George Morton.

We also want to show students and nonstatisticians how to evaluate statistical inputs in their decision making. You do not need a knowledge of statistical theory to deal with many of the misuses which affect decision making in public and private life. While it is common to sneer at the subject of statistics and say that "you can make statistics say anything," it is only through *misuse* that you make statistics say "anything." *Good* statistics are the first step in finding rational answers to important questions!

We hope to show students of statistics that they can learn from the study of misuses. This is an approach that they will not find in most elementary statistics textbooks. However, as Friedman and Friedman tell it:

> Most introductory courses in statistics focus on how to use statistics rather than how to *avoid misusing* statistics. Many textbooks offer a "cookbook" approach to the analysis of data, without advising the reader what will happen to the recipe if one ingredient is left out....
>
> It is our belief that one can learn more from studying misapplications of statistics than from perfect applications. Many disciplines now use case studies as teaching aids. These case studies not only demonstrate the correct way of solving problems, but also show the effects of incorrect decisions, enabling students to learn from others' mistakes. It is one thing to teach students about the assumptions underlying various statistical procedures; it is quite another matter to show realistic examples of the consequences of violating these assumptions [2, pp. 1–2].

This is not a statistical text; we leave to the standard college texts the chore of teaching students the theory and application of statistical tools. However, while not all students of elementary statistics will become statisticians, virtually all students will use and apply statistics in their future professional or private lives. As the statistical theories of the classroom fade with time, the lessons learned from the study of misuses become more important.

In this book, we describe some of the many misuses we have observed and explain why they are misuses. When you have a full understanding of how others create misuses, you are better equipped to avoid creating misuses yourself and to recognize them when they appear in newspaper stories, in politicians' speeches, in advertisers' claims, and even in scholarly journals. Most of the cases cited are real, but we have included several hypothetical cases of misuses committed by Dr. K. Nowall, our imaginary researcher. These hypothetical cases are typical of misuses that have occurred, but we have simplified the situations to give our readers more accessible illustrations.

We also describe some examples of the proper use of statistics. This is not done to take on a function of books on statistical methods but to illuminate our examples of misuses. As the inadequacies of a poorly designed engine can be seen more easily by comparing it to a well-designed one, a misuse of statistics can be more easily understood by comparing it to a proper use.

If it sometimes appears that we are "beating up" on the news media, it is because that is where we find most of our examples—not because the media are more likely to create misuses than others such as corporate public relations people, researchers, government officials, and so forth. Unfortunately, misuses in internal reports or organizational decision making rarely see the light of day. But when a good or bad use of statistics is exposed by the media, it is revealed to all. It is in the media where most of us find out what is happening in the public arena. To this end, it seems to us that the media must be especially careful in their presentations of statistical information.

What Is a Misuse of Statistics?

The following four misuses which follow are the ones we consider most flagrant:

1. Lack of knowledge of the subject matter
2. Faulty, misleading, or imprecise interpretation of the data and results
3. Incorrect or flawed basic data
4. Incorrect or inadequate analytical methodology

The presence of any one of these flaws is enough to create a misuse, and it is not unusual for a number of misuses to occur in a given example. We call this special horror case a *megamisuse.*

Even if the investigator knows the subject and interprets the data and results correctly, the last two flaws (*incorrect data* and *incorrect analytical methods*) can still cause a misuse.

Correct and *incorrect* are absolute terms, but in evaluating methods and data we frequently must assign relative measures of correctness. For example, you might think that getting data for the number of accidents during a particular time period on a particular stretch of highway would be a simple counting process. But it can be quite difficult to count accidents in such situations (see Chapters 4 and 5). Despite your best efforts, you may be in the unpleasant situation of having to admit that you may be correct or incorrect. In this case, all you can say and not commit a misuse is that, *within a particular context,* the data are "reasonably correct," or "probably incorrect," or some similar relative statement. Why?

A multitude of errors can be responsible for a misuse or megamisuse of statistics. The people who do the actual counting can make errors (have *you* ever been distracted when counting money or measuring ingredients?), workers can lose reports at any stage, recorders of events can give incorrect dates, different observers may have different definitions of what an "accident" is, computer malfunctions may affect the data, computer programs may perform improper functions (unknown to the users), and so forth. The investigator has an obligation to report all the known sources of flaws in the data and to hypothesize likely unknown sources. In addition, to maintain credibility and assist the readers in their evaluation of the results, the investigator should discuss the possible effects (or noneffects) of data errors.

In analyzing data, more is not necessarily better. Unfortunately, it is not always possible to have one uniquely correct procedure for analyzing a given data set. An investigator may use several different methods of statistical analysis on a data set. Furthermore, different outcomes may result from the use of different analytical methods. If more than one conclusion results, then an investigator is committing a misuse of statistics *unless the investigator shows and reconciles all the results*. If the investigator shows only one conclusion or interpretation, ignoring the alternative procedure(s), the work is a misuse of statistics.

Sometimes you will find it hard to make a clear distinction between the proper and improper use of statistics. Political polling is a good example. The interviewers can ask respondents how they feel about candidate X and many respondents will say that they like candidate X. Then the interviewers ask whether the respondent will vote for a particular candidate. A large proportion say that they "do not know." Of those who say they will vote for a particular candidate, X gets the largest proportion. But what is that proportion?. Let us say the results are:

Candidate X:	40%
Candidate Y:	24%
Don't know:	36%

What can you do with these results? You could conclude that candidate X "leads the pack" (40%). Conversely, you could conclude that a majority— 50%—(the Don't knows (36%) plus those who favor candidate Y (24%)) do not favor X but for reasons known only to themselves, 36% refuse to publicly acknowledge their intent to vote for Y.

Here, we show two possible interpretations of these results, neither of which can be shown to be superior to the other in the absence of additional information. Which interpretation is correct? The data and analytical procedures of this survey may be correct, but the results are inconclusive; only the election will provide the answer. If the report gives only one interpretation, then it is a misuse of statistics; if both or other likely interpretations are reported, then we have a proper use of statistics.

SOME CONSEQUENCES OF MISUSES

The erroneous results of the misuse of statistical analysis can be slower to surface than errors in other types of analysis. For example, in civil engineering, a missing fastener might result in an immediate and verifiable failure when the bridge collapses! But in setting public policy, a statistical misuse can lead to wrong decisions with consequences that may not be obvious for decades. In Brazil, for example, a mistaken belief that "street children" (children who apparently were homeless and lived on the streets) numbered in the millions, led to a lack of action by government and social agencies who thought the problem insoluble. But when a relatively precise count was actually taken using a consistent definition of "homeless," the investigators found that true street children numbered only in tens of thousands. The public saw this as a reasonable number for public action, and supported corrective measures.

Because technological misuses of statistics are, hopefully, self-identifying and often do not appear in print as do statistical statements pertaining to other areas, most of our examples of misuses are drawn from nontechnical situations, as in the following.

An Unintelligent Misuse to Support Racism

Racism is a persistent problem in human societies. An early misuse of statistics in support of racist views has persisted from the early 1900s to the 1980s. This argument for the inferiority of certain presumed genetic groups said that the genetically unintelligent quit school young. Many believed—and still believe—that intelligence quotient (IQ) test results justify this argument. Stephen Jay Gould recently studied this belief:

> Of all invalid notions in the long history of eugenics—the attempt to "improve" human qualities by selective breeding—no argument strikes me as more silly or self-serving than the attempt to infer people's intrinsic, genetically based "intelligence" from the number of years they attended school. . . . The genetic argument was quite popular from the origin of IQ testing early in our century until the mid-20s, but I can find scarcely any reference to it thereafter [3].

Gould shows that subsequent statistical analyses reveal little if any relationship between years of schooling and intelligence. More years of schooling do not necessarily indicate more intelligence.

Watch Out for the Curve Ball

Four hundred thousand copies of *The Bell Curve: Intelligence and Class Structure in American Life*, by Richard J. Herrenstein and Charles Murray, were in print two months after its 1994 publication. It is something of a mystery to us whether so many readers actually plowed through the 800 pages of text and the abundance of data and statistical analysis. Unfortunately, whether they read a part or the whole of the book, the readers were witnesses to one of the premier statistical megamisuses of our time.

The heart of *The Bell Curve* is a regression analysis of data on 12,000 subjects taken from the National Longitudinal Study of Youth (NLSY). From this analysis and the referenced works of others, the authors argue that genetic differences in intelligence exist among races. In particular, they state that blacks underperform Caucasians in IQ tests and that the many documented socioeconomic problems of American blacks are caused by this genetic deficiency in IQ or intelligence.

The catalog of statistical misuses produced by Herrenstein and Murray, as revealed by the many distinguished scientists and authors who reviewed the book, is so long, and the statistical analyses are so fundamentally wrong, that even a superficial review invalidates the authors' arguments. Of prime importance is their definition of intelligence as a single number, the score on the standardized Stanford-Binet test. This measure, the IQ score, is a statistical artifact, an issue of definition which we discuss in Chapter 4.

Unfortunately, that pseudoscientific megamisuse became a best seller. Did it achieve its commercial success because it appeared to give scientific support to those who are searching for a scientific justification of their own prejudices?

DELIBERATE OR INADVERTENT?

Are we surprised when misuses and errors support the viewpoint of the originator? President Reagan (1981–1989) overestimated the growth in work time needed to pay taxes by a factor of almost eight and the growth in the percentage of earnings taken by the federal government in taxes by more than 10 times (4). Similarly, Samuel George Morton, who had the world's largest collection of pre-Darwinian skulls, manipulated the data to support his conclusions about the inferiority of non-Caucasians (5).

Is There a Smoking Gun?

The analyst must be careful in assigning motives, for we are all prone to make errors in our favor. It is tempting to say both of the above examples were deliberate misuses of statistics and to attribute conspiratorial intent to the originators. However, without having the smoking gun—explicit statements of the intent to deceive—you can't know a person's motives.

President Reagan's incorrect statistics may have come from "number illiteracy" rather than deliberate misstatement or an intent to mislead. As an economist noted at the time, "For Mr. Reagan, it is not just a question of numbers, *which are not his strong point* [italics ours] . . ." (6). Samuel George Morton published all his raw data; thus, we can show his manipulations of the data and derive the correct results. Because he did not attempt to hide the data, we may conclude that he did not intend to mislead his readers and probably did not realize that he was manipulating data to fit his preconceived ideas.

NO LITERARY LICENSE IS GRANTED TO MISUSERS OF STATISTICS

We have seen a case in which the originator of a misuse admits that it is deliberate. In their article in *Science,* Pollack and Weiss stated that "the cost of a telephone call has decreased by 12,000%" since the formation of the Communication Satellite Corporation (7). Letters to the editor of *Science* pointed out this misuse of statistics. Pollack replied: "Referring to cost as having 'decreased by 12,000%,' *we took literary license to dramatize the cost reduction* [italics ours]" (8). If you want to avoid dramatic statistical misuses, don't confuse literary fiction with statistical fact. In Chapter 12, we give more examples of misuses in involving percentages.

Play Fair When You Play Statistics

You have the right to suspect a deliberate intention to mislead when someone refuses to supply the raw data and/or sources for the data. If the report supplies sources of data or the data itself, then you have no reasonable basis to suspect deliberate deception. There are occasions, however, when some "think tank" issues a paper and there is so much information, much of it not attributed, that the lone statistician seeking truth in the conclusions would have almost no chance of refuting statements because of the time it would take to prove or disprove the conclusions. But even in these cases:

> Gross flouting of procedure and conscious fraud may often be detected, but unconscious finagling by sincere seekers of objectivity may be refractory. . . . I propose no cure for the problem of finagling: indeed, I . . . argue that it is not a disease. The only palliations I know are vigilance and scrutiny (9).

2
Setting the Stage:
Categories of Misuses

I must create my own System or be enslav'd by another Man's.
—Blake

Our little systems have their day;
They have their day and cease to be.
—Tennyson

INTRODUCTION

We see three common denominators in all misuses. First, the writer or investigator does not understand the data that are used or reported. Second, the writer or researcher does not understand the quantitative methodology. The methodology may be simple or complex; without understanding of how it was devised or when particular techniques are appropriate, misuse results. Third, the writer or investigator fails to proceed logically from purpose to measurement to analysis to conclusions.

Because similar offenses often occur within categories, we have set up subcategories to simplify identification. In addition, many examples of misuses which involve several offenses we call *megamisuses* are also included.

In the first edition we organized statistical misuses into five major categories, which was adequate for our purposes at that time. In this edition, we have added a new category which comes from accumulated experience, entitled "Thinking."

Here are our six categories of statistical misuses and their subcategories:

1. What you don't know *can* hurt you: lack of knowledge of subject matter.

2. You can't make an omelet out of confetti: Poor quality of the basic data
   Blatant flaws
   Bias, in any of its many forms
   Bad definitions: inadequate, inappropriate, improper
3. All's well if it ends well: faulty studies
   Misguided design of the study and report
   Visually deceptive graphic presentation and what can result
   Faulty interpretation of findings
   Misleading presentation of results
4. Use the right tool for the job or take the consequences: incorrect or incorrectly applied statistical methodology
   Choice of the wrong method
   Erroneous calculations
   Pitfalls in regression analysis
   Thoughtless sampling
   Overlooked effects of variability
   Failure to verify
5. A knowledge of statistics is not a substitute for thought: lack of thinking
   Failure to recognize absurdity
   Missing the obvious
6. Sooner or later the truth will out: deliberate suppression of data

In this chapter, we set the stage for the balance of the book with brief discussions of each of these categories. In the following chapters, we present more detailed examples of each type of misuse, explain why we consider them to be misuses, identify some of the people who originate misuses, describe some of the statistical methods which are misused, and show the dangers of misuse. We also discuss Ockham's Razor, which relates to many types of misuses. In addition, we give considerable coverage to the creation of mythical numbers which we call "ectoplastistics," and the actions of past and present governments in misusing statistics.

A CLOSER LOOK AT THE CATEGORIES OF MISUSES

Lack of Knowledge of Subject Matter

Sometimes researchers or writers do not understand the basic data with which they are working and do not know how to compute a statistic, how to formulate a hypothesis, how to get existing information, or how to test the results for validity (the connection between the concept and the measured value). Thus we have the spokesperson for U.S. Representative J. Dennis Hastert

(R-IL) speaking of the representative's opposition to using statistical sampling in the year 2000 census, saying, "We saw where sampling was used in the past—it was statistical guessing." Perhaps the representative is unique among politicians and does not use polls—a form of statistical sampling—but most of his colleagues do and they rely on their pollsters not only at election time but during their time in office when voting on different issues. In the case of opinion polls, however, there can be differences between the results and what actually occurs. This is because the pollsters are sampling opinions, which are subjective in nature and open to the possibility that the respondents will change their minds. But much statistical sampling is not guesswork, but scientific work, as manufacturers of drugs, appliances, automobiles, and fighter planes can testify. Does the congressman plan to stop driving or trust to nature the next time he gets ill?

Poor Quality of the Basic Data

Blatant Flaws

Sometimes the basic data are not available or cannot be validated. Makers of public policy may use statistical analyses based on data obtained in situations where the statistics presented cannot be verified. Some recently published examples of such data are the number of deaths from starvation in a remote area, the number of people killed in a genocide during a civil war, the number of illegal aliens living in the United States, and the number of women who are the victims of violence in Connecticut each year—flaky estimates one and all.

In other cases, the facts are known but are *misrepresented* and the person doing the analysis is unaware of this misrepresentation.

The data may not be *comparable,* as when comparing certain test scores for public school students in different states. (Different tests or different combinations of tests may have been used.)

Data may be *invalid* for the purpose at hand. Is your concern the ratio of total taxes to total profits? Then the ratio of U.S. taxes paid to the total profits of a multinational corporation (which includes overseas earnings) is not a valid measure, since taxes may have been paid in foreign countries (1).

Some data are *invalid* under any circumstances because of measurement error. Measurement errors plague the polling analyst as much as the engineer and physical scientist. For example, what kind of data can we expect to result from a misconceived poll question such as, "How many hours do you commute to work each day?" Without specifying whether commute time is one-way or round-trip, we can be sure that there will be extreme measurement errors, invalidating the results.

Bias

Data may be biased because of the manner of collection. Leading questions in questionnaires can be as subtle as a sledge hammer. A "national referendum" on population policy asks: "Yes or No: Before reading this mailing, were you aware of the gravity of our national and global population crisis?" This question *assumes* that there is such a crisis. There are many who disagree with this assumption. However, if someone who disagrees answers the question as posed, a "no" answer makes that respondent a party to the assumption that there is a crisis and that it is a grave one (2).

In a questionnaire designed as part of a research project to detect bias, the same respondents will oppose a constitutional amendment to prohibit abortion but support a constitutional amendment to protect the life of the unborn child (3). In this case, a question designed to determine an underlying attitude produces a different response depending on the wording of the question. Do you think these different answers represent different attitudes?

Bad Definitions

Many countries are concerned about rising numbers of "homeless people." But what *are* those numbers? You must know what a homeless person is in order to add him or her to the count. Pedestrians and police may think that a homeless person is someone who is sleeping on the sidewalk, but he may well have a home to which he does not choose to go. U.S. federal statutes define a homeless person as a person who "resides in grossly inadequate or grossly substandard accommodations" (4). Are persons homeless if they are living with relatives on a temporary basis, as some investigators say? Do we define homelessness by behavior (sleeping on streets, living with relatives, living in inadequate housing) or by status (having no legal claim on a home, or no income)? As Rep. David Dreier said in a Letter to the Editor printed in the *Wall Street Journal*, "We may never be able to deal with the homeless problem until there is a consensus on what it means to be 'homeless'" (5). One thing is certain, any measurement of the homeless can make sense only in the context of a valid, meaningful, and consistent definition.

Faulty Studies: Preparation of the Study and the Report

Design: A Tooth for a Tooth

If an experiment is to be valid, it must be designed to give clear and identifiable results. To introduce the important and complex subject of the design of experiments, we examine a real-life experiment.

Almost all of us have or have had wisdom teeth, so we base our example of experimental design on a research paper which was concerned with testing a new method for reducing infection in the holes ("sockets") left when wisdom teeth are removed (6, as discussed in 7). In this example, the reduction in infection was to be measured by comparing the new method to an existing treatment. But individuals vary greatly in their potential for infection, and their response to medication depends on their age, health, inherent characteristics of the immune system, and other confounding factors.

How are we to separate out these confounding factors? One way would be to simultaneously extract a tooth from each of a pair of identical twins. But it is unlikely that both would need an extraction at the same time. Although it would be a contribution to dental science, to perform extractions for no other reason than to complete the experiment is unethical. However, in a reasonable time period we might be able to find two groups of individuals needing extractions who are similar in most of the confounding factors we can conceive of. (We might even be able to find individuals who need two extractions at one time.) The group receiving the new treatment is the *experimental group* and the group getting the existing treatment is the *control group*. The control set is composed of the sockets (after extraction of the tooth) in each mouth that get the existing treatment; the experimental set is composed of the tooth sockets receiving the new treatment.

How do we assign the new treatment to the different teeth to assure that there is no systematic effect? After all, there *might* be a difference in response among many humans between right and left sides of the mouth. We do not say there is, but experiments with other mammalian characteristics have shown differences between sides. To reduce the effect of this factor, we "randomize" the assignment of treatment to side of the mouth, using some random process. Once we would have considered tossing dice or using a random number table. Today, however, we are most likely to use random assignment by computer.

How do we make sure that the dental surgeons apply the new and existing treatments so they do not influence the results, consciously or unconsciously? Supervision is one approach, but in general it is better to make a "blind" experiment in which the surgeon does not know which treatment he is applying, the new or the old.

Alas, determining the intensity of an infection is largely subjective, a determination made by the surgeon with support from the patient plus some objective measurements. How are we to reduce any conscious or unconscious influence on the results in this case? Once again, we can use a blind approach, with both the surgeon making the evaluation and the patient unaware of which tooth has been exposed to which treatment. When both the application of the trial method and the evaluation are performed blind, we call it a *double-blind* experiment.

This example is a brief, and simplified, introduction to some of the pitfalls of experimental design and their resolutions in experiments. The public (some of whom may have wisdom teeth removed some day!) has a right to expect that a professional researcher will know the principles of experimental design and will use them appropriately in experiments.

Despite the fact that professionals (and even many nonprofessionals) know these principles, scholarly papers of greater and lesser importance violating one or more of these principles still appear in refereed scientific journals. Sometimes the conclusions published in scientific papers are reported in the media, and, unfortunately, these media reports seldom include information on the design of the study so that the concerned reader can make an informed judgment. To their credit, some media do, on occasion, succinctly communicate those design facts that are important to the nonprofessional reader or listener, as *The New York Times* does when reporting its survey results.

Graphic Presentation

Graphic misrepresentation is a frequent misuse in presentations to the non-professional. The granddaddy of all graphical offenses is to omit the zero on the vertical axis (as we discuss in detail in Chapter 6). As a consequence, the chart is often interpreted as if its bottom axis were zero, even though it may be far removed. This can lead to attention-getting headlines about "a soar" or "a dramatic rise (or fall)." A modest, and possibly insignificant, change is amplified into a disastrous or inspirational trend.

Interpretation of the Findings

How many are the ways to misinterpret findings! How easy it is to create an exciting headline by projecting to the population as a whole results from a sample that was selectively, not randomly, drawn. For example, when many academic researchers wish to test a hypothesis, they find that the students in their classes are a ready source of subjects. The researcher-teacher administers a survey to a small group of students. If the researcher doesn't project the results to the whole population of the United States, some eager or uninformed headline writer will.

Americans have shown a continuing public concern for education in the United States which dates back to before the writing of the Declaration of Independence: "In 1749, Benjamin Franklin, who was himself largely self-educated, published a pamphlet entitled *Proposals Relating to the Education of Youth in Pensilvania* [sic], which reads like the prospectus for a progressive school of the twentieth century" (8). Since before Franklin's time, debates about education have been frequent public events, as is the presentation of statistics to prove a point on one side or other (or on both sides). In modern times, test scores have proved to be an irresistible temptation for the debater, for they *appear* to be an objective measure of performance.

In 1989, the Congressional Budget Office, in its report, "Educational Achievement: Explanations and Implications of Recent Trends," warns about "how standardized achievement tests are used and *misused* [our emphasis] in policy discussions" (9). This is nothing new! Politicians, educators, and the public alike often take high school student performance on the Scholastic Achievement Test (SAT) as a measure of the quality of education received by the student. In 1974, a school introduced a competency program and then showed a gain of 31% in the SAT pass rate over the pass rate of 1971. In a classical misuse of statistics, the school's administrators claimed that this increase was a result of the program. But the administrators did *not* report that the rate had already risen considerably in 1973, one year before the program was introduced (10). Debaters intent on furthering one or another educational policy perpetuate this kind of misinterpretation. In Chapter 12 we describe another take on national educational testing.

Presentation of Results

Results are one thing; the way they are presented is another, especially if essential elements are omitted. Results are often presented as being significant without any indication of the sample size (from which the sampling error can be deduced).

It is good practice for a pollster to include an appropriate response— "No Opinion," "Don't Know," or "Do Not Choose to Answer"—in the questionnaire. If the report of the poll does not give the proportion of these responses, it is a clear misuse by misinterpretation. We see too many reports in which it is claimed, "Twice as many in favor as against," and which do not reveal that just as many respondents did not choose to answer or had no opinion. When these categories of nonresponse are reported in the description of the results, the reader's conclusion is likely to be different from the one illustrated in the advertisement or article.

New York Times reporter Malcolm W. Browne correctly observes that "A lot of people are afraid of numbers, so numbers make wonderful cudgels for winning arguments" (11).

Browne goes on to describe a television commercial in which the narrator tells the audience that his product is superior because it contains 850 milligrams of "pain reliever" while a competitive product has only 650 milligrams. Questions which arise are: Is the "pain reliever" the same in both products? If not, which pain reliever is best for a particular case? If they are the same, is this pain reliever efficacious? Is it better to take more than 650 milligrams in one dose? By not dealing with such questions but simply repeating "850 milligrams," the advertisement misuses numbers—statistics —to create an impression that may be incorrect. Despite a number of legal and regulatory actions, you can still see such misuses in advertising. Deliberate? Or is it possible that the ad writers just don't know the errors of their ways?

Statistical Methodology

General

There is no shortage of statistical methods. Elementary statistics textbooks list dozens, and statisticians constantly develop and report new ones. But if a researcher uses the wrong method, a clear misuse, to analyze a specific set of data, then the results may be incorrect. For example, if a researcher uses the "wrong" *typical value* (average), this may lead the reader (or the researcher!) to a wrong conclusion. There are several ways to identify a "typical value": arithmetic mean, median, mode, geometric mean, harmonic mean, midrange, and others. Each has its own computational formula, and, to be correctly used, each must be suited to the particular data and purposes of the analyst. To determine the "typical" salary in a community where a few individuals have extremely high salaries, the knowledgeable analyst uses the median, which is more representative of the salaries earned by people in the community than the arithmetic mean. The use of the mean where the median is appropriate creates a false impression (see Chapter 7).

Erroneous Calculations

Neither careful proofreading of newspaper copy nor the referee process for academic journals guarantees error-free computation. We see many examples of incorrect results in computations involving only elementary arithmetic and have published, alas, such errors ourselves. In a *Wall Street Journal* article on the effect of heart surgery, we found four incorrect results from faulty addition. None of these errors affect the conclusions, but (surprise!) the errors emphasize the desired effect (12). In many cases, this may not harm anyone, but computational errors in a life-or-death situation could prove otherwise. Will these errors be avoided now that the personal computer has replaced the abacus? Not unless the analysts or reporters make the effort to reconcile totals. As we show in Chapter 12, a reporter who is asleep at the switch can make an absolutely absurd error. It's easy enough to check using rough manual or calculator computations, but you must *think* to do it.

The use of computers does not necessarily mean a reduction in errors. It may even make things worse, possibly much worse. While the computer may make fewer arithmetic errors, it can create new kinds of data-related errors which can be even harder to find.

A team of researchers in the field of aging said that increasing numbers of the elderly moved to metropolitan areas in the last decade. They used data on computer tapes supplied by the U.S. Census Bureau to arrive at this conclusion. But there was an error in the data which:

> . . . involved a change in the way the bureau arranged the information on the computer tapes of its 1980 data. . . . In previous censuses, . . . those

who moved to the United States from abroad were not counted as part of interstate migration within America.

In 1980, . . . the census counted 173,000 people who came to the United States from abroad during the last decade and "lumped them into the group" of elderly people moving from nonmetropolitan areas. . . . Thus those who had moved from foreign countries were counted as part of interstate migration. . . . The researchers failed to notice that the two kinds of migration had been lumped together [13].

Does the Social Security system contribute to a reduction in savings and thus a reduction in investment capital? A computer programming error resulting in erroneous calculations led to a multibillion-dollar overestimate of the negative effect of Social Security on national saving (14). All that computer power was brought to bear, and it produced a false result. Interestingly, the director of the project, while acknowledging the error, stated that he had not changed his beliefs.

Let's not trouble him with facts.

Regression

In a few words, *simple linear regression* is a systematic process for drawing a straight line through a series of data points in a two-dimensional plot. (Figure 2.1 shows the two-dimensional plot of data points and the regression line

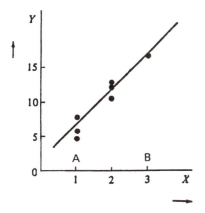

Figure 2.1   An example of a *regression line*. The black dots are the observed data points having particular values of the variables on the X- and Y-axes; the line is obtained by "fitting" it to the observed data points. A common use of the regression line is to predict values of Y based on knowing a value of X. When the value of X lies outside the range of observed points (as indicated by "A" and "B"), you cannot be sure the relationship holds.

drawn through them.) This is a potentially powerful statistical methodology, for once you have drawn a *regression line* you can predict the value of Y (along the vertical axis) by knowing the value of X (along the horizontal axis). Whether or not the prediction is valid is another story, which we discuss later.

But there is no gain without pain. Along with the many benefits of regression, we accumulate new kinds of misuses. Even nonstatisticians readily recognize some of the misuses of regression. For example, you are right to suspect predictions of values of Y based on values on X that are outside the range of the data from which the line was obtained (lower than 1 or higher than 3 in Figure 2.1). But if you have only the equation of the line (the "regression model") and don't see the plot of the points, then you may not realize that you are using a value of X outside the range of observed values.

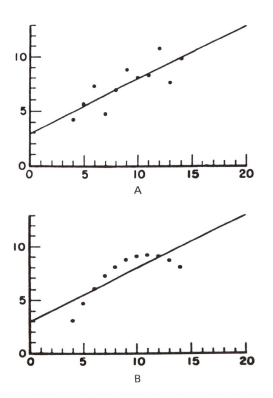

Figure 2.2 Two dramatically different sets of data are shown with the fitted regression lines. The lines are identical! Imagine what a misuse of statistics it would be to use the line in (A) to project a value on the Y-axis for a value of 20 on the X-axis. (*Source*: F. Anscombe, Graphs in statistical analysis, *The American Statistician*, 27(1):17 (1973). Used by permission. All rights reserved.

Linear regression, the form usually used, gives you the formula for a straight line and some statistics on how well the line "fits" the data. If you rely on such summary measures and do not look at the plot of the points, then you can easily fail to see important relationships. In Figure 2.2, we show two dramatically different plots of points. Despite their obvious gross differences, the regression fits the same line to both of them. The equations you get from a computer for these two lines are identical. Don't get caught in this trap! Always look at the plot of the points before drawing a conclusion.

Sampling

Redbook magazine surveyed its readership on sexual attitudes using a questionnaire inserted in the magazine. Of the more than 3 million subscribers, 20,000 women and 6000 men selected themselves and responded. Why did they select themselves? Were the reasons such that these respondents form a special group? Would you project the responses from this self-chosen group to the entire population of the United States? Although the *Redbook* article was exemplary in cautioning readers that the sampling (only part of the whole population responded) was not random and could not be projected to the whole population of subscribers, the *New York Post* article describing the results went even further and did not caution its readers about the errors made by projecting the results to the *general* population (15).

Modern Maturity also entered the sexual attitude sweepstakes, reporting on the work of Beverly Johnson, assistant professor of nursing at the University of Vermont. Professor Johnson announced that her study of 164 men and women between ages 55 and 60 "should lay to rest the three most common stereotypes about older people," implicitly projecting her findings to all Americans in that age group (16). She conducted her "ongoing research" through the use of the responses of people who contacted her for a copy of her questionnaire and then *chose* to submit it. This practice continues unabated—aided and abetted by modern technology.

Questions are put before the public by radio, television, or in print media and the audience is invited to respond, using special telephone numbers. For a modest cost the callers can give their response and become part of the total results to be announced at a later date. As one cynical observer noted, "It looks like a poll, it sounds like a poll, but it isn't a poll" (17).

Charges of sexual abuse in New York child care centers were the impetus for attempts to estimate the number of employees at the centers who had criminal records. The then Mayor Edward I. Koch released the results of an initial sample of 82 employees in which about 44% of the employees checked reportedly had criminal convictions. The headlines were dramatic. Subsequently the city's Human Resources Commissioner, George Gross, reported on an analysis of a sample of 6742 fingerprints of employees at child care centers and showed that 253, or about 4%, were found to have criminal records.

The initial criticisms were directed to the the small sample size (82) of the first report: "Mayor Koch released those first figures (from the sample of 82 employees), despite the objections of some mayoral advisors who were concerned that the sampling was too small to be valid . . ." (18). However, Commissioner Gross felt that the second sample of 6742 employees was more than adequate: "With this kind of huge sample, my feeling is that the numbers will probably hold up . . ." (18).

The real issue is that neither Gross nor Koch dealt with *how* the samples were selected. Were the first 82 sample employees chosen randomly from the population of about 20,000 such employees in the city? If they were, then it is reasonable to project that between 33% and 55% of the employees had criminal records. If the true proportion of employees with criminal records is below 5%, as shown by the second, and larger, survey, then it is extremely unlikely—almost impossible—to have obtained the results cited by the mayor, *if the sample was random.* And if it was not random, it was a misuse of statistics for the mayor or the media to make projections to the population of all child care center employees. Thus, we have no basis for an estimate of what proportion of the employees had criminal records.

On the other hand, the city's Human Resources Commissioner gives no information as to how *his* sample of 6742 was chosen. We can draw no meaningful conclusions about either estimate from the results as stated, because we do not know how these samples were chosen.

Variability

When survey results are given as proportions and without mention of the sample size, there is no way to estimate the sampling error—that error in the results which is due to the chance effects of sampling. Are you surprised that the advertisement does not tell the sample size when it states that "9 out of 10 MERIT smokers (are) not considering other brands" (19)?

You have good cause to be concerned when two medical researchers give several percentages concerning vasectomy and cancer of the cervix without giving the sample size (20). In fact, you have even more good cause to be concerned when you also factor in the carelessness of the editors of the journal and/or media reporters who failed to query the researchers about the missing sample size.

Has the jobless rate risen when the headline says: "Jobless rate in US up slightly . . . to 5.7% from 5.6%" (21)? You need a knowledge of the sampling error (variability) to answer this question. In this case, although the headline was misleading, the author of the article knew enough about the subject to state that the jobless rate had held "essentially constant." But if the author of an article reporting a statistical result is reluctant to make the comparison between the change and the observed variability, then the reporter should give some measure so that the readers can make their own evaluation. In Chapter 9 we show that you can easily evaluate a survey when you have been given enough information about sample size and the method of selection. A few newspapers, including *The Washington Post* and *The New York Times*, routinely include, as a part of every survey, both the necessary information and the results of the calculations.

Verifiability

Statistical results often are held in such low esteem that it is imperative that the results be verified. Unfortunately, we often cannot confirm the results obtained by authors. The reasons are many: computations on the data given do not yield the same results as those reported; insufficient data are given, so that readers cannot check the results; there is insufficient information on how the sample was chosen; the sources of data are not stated; or the author has collected and processed the data so that it is impractical, if not impossible, for anyone else to check the reported results.

Sometimes, the use of a computer can muddle any effort to verify a result. A group of physicists, computer scientists, and others complained that some computer programs were too complicated and thus so difficult to read that a reader seeking to verify the results could not determine exactly what was done to the data. This group is now reluctant to accept the results given by researchers who make extensive use of this type of computer analysis (22).

Lack of Thinking

We are concerned with *thinking* about statistics, the process by which we direct our studies to the right measurements for our purpose and make sense of statistical results and give sensible descriptions of them. Correct thinking about percentages eludes the layperson, the scientist, and the journalist alike. Fallacious and inadequate thought can lead to decisions with dramatic consequences in medicine and courtroom processes. In public policy debate, the failure to move from concept to a valid measure of that concept leads to deceptive and confusing conclusions and bad decisions. All this and more, we make clear for you in Chapter 12.

Deliberate Suppression of Data

As we discussed in Chapter 1, it is hard to know what is in someone's mind when that person takes an action. What appears to be a deliberate action may in fact be an error or an act which the research investigator sincerely believes to be correct and appropriate—then again, it *might* be deliberate.

SUMMARY

Has the misuse and abuse of statistics declined or increased with the increase in the proportion of the population with higher education? Has the widespread use of personal computers reduced or added to the number of misuses? Has the Internet helped to identify misuses by making data instantly available at any location in the world, or has it multiplied misuses? We don't know the answers to these questions, but we can show examples that help to identify those misuses that do occur. We all need a basis for recognizing and dealing with misuses of statistics.

Methods of mass communication—which include the media and the less structured Internet—have gained greatly in speed and access in the past few years. We have the instant access of television reporting with its (to some degree understandable) lack of depth in coverage of complex topics; the interaction between the observed and the observer in TV "surveys"; and the effect of statistical projections on the behavior of voters whose behavior is being projected. The Internet brings even faster access to a worldwide audience, often with a considerable depth of coverage, incorporating text, audio, and video into one medium. But some of the information now available on the Internet can also be inaccurate, fantastical, or purposely misleading, and the casual or naïve reader must be wary!

Sample surveys and polls are a cultural institution and have great influence. Often, misusers incorporate the full spectrum of misuses described in this book into the published results. The public, the policy makers, and those who report on the surveys and polls appearing in the media must understand the limitations of statistical methods and how statistics can be misused, or suffer the consequences of poor decision making.

Vigilance and *scrutiny* are the hallmarks of the informed researcher, reporter, and reader.

3
Know the Subject Matter

Not well understood, as good not known.
　　　　　　　　　　　　　　　　　—Milton

Knowledge is of two kinds; we know a subject ourselves, or we know where we can find information upon it.
　　　　　　　　　　　　　　　　　—Samuel Johnson

INTRODUCTION

To avoid a damaging, possibly dangerous (or even embarrassing) misuse of statistics when analyzing data on real-world problems, the writer or researcher must be knowledgeable not only about statistical methodology but, more importantly, about the subject matter.

Suppose there are conflicting sets of statistical data, some good, some bad, and some useful to a limited extent, or some or all of your sources of data are flawed or inappropriate. If you do not know the subject matter, you may fail to recognize this fact. Thus, even though you use appropriate statistical tools and make no arithmetic errors, if you are not knowledgeable about the subject you are investigating, the results might well be incorrect or flawed. *Only* if you know enough about the subject can you sort the good from the bad and avoid a misuse of statistics.

DR. K. NOWALL AT WORK

Unexpected Prospering of the Native Americans

Dr. K. Nowall, a scholar who never heard the adage "Better be ignorant of a matter than half know it" (Pulilius Syrus), was amazed when he discovered the nearly sixfold increase in the Native American* population from 1860 to

---

*This category includes the American Indian, the Eskimo, and the Aleut.

1890. He found that the U.S. Census counts of this minority population were 44,000 in the 1860 decennial census report and 248,000 in the 1890 report (1). "This may be the most rapid rate of population growth ever seen, a compound average growth rate of almost 6% per year. That's phenomenal!" said our ambitious scholar.

Unfortunately, Dr. Nowall did not know the subject matter. If he had known the demography of Native Americans, he would have known that in 1860 most Native Americans lived in Indian Territory or on reservations (lands set aside for these peoples) and they were not included in the U.S. Census counts. "American Indians were first enumerated as a separate group in the 1860 census. However, Indians in Indian Territory or on reservations were not included in the official count of the United States until 1890" (1). Since Dr. Nowall did not know that this previously excluded group was now *included* in the 1890 census, he saw a "soar" where more knowledgeable people saw a natural rise in the reporting, not the count.

SOME REAL-WORLD EXAMPLES

Using the Space Shuttle to Cross the Mississippi

A researcher working with computer graphics who "discovered" a statistical anomaly gives us another example of what happens when inadequate knowledge of the subject matter and carelessness can lead to an unfortunate waste of time and talent (2). The computer researcher and his co-workers, anxious to show clearly some data on motor vehicle fuel consumption, used already published data to create a series of complex computer graphic displays. Using all the latest statistical tools, they determined that there had been a change in the definition of motor vehicle horsepower during the period when the data were generated.

But persons familiar with the subject matter already knew that a change in a critical definition had occurred. In fact, the very published data which were used by the researchers to create the displays from which they deduced the change in horsepower definition *had a footnote which clearly stated that the definition had been changed*! They had used computing power and sophisticated software to discover a fact that was known to most auto mechanics. This is a case where the lack of knowledge of the subject led competent investigators to violate Ockham's Law, which we discuss in Chapter 11.

Judge Not, In That Ye Be Judged Yourself

When we wrote up the following example in 1987, we felt that the newspaper story in 1979 claiming that women in the federal judiciary were underrepresented had not taken into consideration the fact that not many women had received JD's 10 to 15 years previously. This is how we phrased the example at that time:

> Only 28 of the nation's 605 federal judges are women even though more than 45,000 women are now practicing law in the United States [3]. The writer claimed that, since about 10% of the 478,000 lawyers in 1979 were women, at least 10% of the 605 federal judges should be female. Since there were only 28 women federal judges, or about 5%, women were "obviously underrepresented."
>
> But a proper evaluation of the claim of underrepresentation is based on the answer to this question: How much time elapses between getting a law degree and becoming a judge? Federal judges are not appointed from the ranks of new law school graduates. From the difference of 13 years in the median ages of lawyers and judges [4, Table 3], we conclude that it is reasonable to assume that, on the average, it is 10 to 15 years before a newly-graduated lawyer enters the "eligibility pool" for federal judgeships.
>
> Thus women lawyers eligible for a federal judgeship in 1979 would have had to receive their law degrees prior to 1970. What was the proportion of law degrees awarded to women prior to 1970? Never greater than 4% [5, p. 280]. If you compare this 'eligible pool' of women lawyers to the 4.6% of federal judges reported to be women, there is no basis to conclude that, in 1979, women were underrepresented in the federal judiciary.

Were we right? Was it just a matter of time before women lawyers were better represented on the federal judiciary? Have things improved in the 10 years since we wrote the above? Not by enough, it would seem. In 1980, which represents the part of the pool from which judges today might be drawn, 34,590 JD's were awarded. Of this number, 10,754, or 31%, were women. In 1996, about 27% of the lawyers practicing were women, and there were 140 female federal judges out of a total of 842 federal or a total of 17% (6,7). Things are looking better, but if women lawyers were appointed federal judges in proportion to their numbers, there would be 227 female federal judges, not 140. The scales of justice are still tipped in favor of the males. Isn't it about time the politicians (and voters) took notice and asked if the scales of justices are fairly balanced?

The Sky Is Falling (Again)

At the end of the summer of 1997, when a booming economy and bipartisan budgetary agreements made it difficult to make headlines proclaiming the imminent demise of Medicare and Social Security, Senator Grassley of Iowa found a new cause for public panic—pension plan underpayments. Senator Grassley, Chair of the Senate Committee on Aging, announced that at least

8% and possibly as much as 20% of the 14 million pension-drawing retirees were getting smaller payments than they were entitled to and that "everyone was at risk" (8).

The source of this distressing news was data supplied to the committee by a government agency, the Pension Benefit Guaranty Corporation (PBGC). The PBGC pays retirement benefits to pensioners of those companies that cannot pay the promised benefits because their pension plans are inadequately funded. The PBGC reported to the committee on the results of an audit of 6000 pension plans that were closed in 1994 and 1995 and which distributed the assets as lump-sum payments to workers. In this audit, they found an 8% error rate.

Unfortunately, the committee chairman, whose government pension is not subject to termination and consequent lump-sum payout, apparently confused the two types of payment—those that are paid out regularly to retired workers from fully funded pension plans, and those that are paid out as lump sums from underfunded plans. This led to the headline-worthy statement that *everyone* was at risk for their pension payments.

No federal agency, including PBGC, has audited the regular monthly payments that are received by most retirees. However, spot checks have shown these payments to be generally correct, and certainly not anything that would suggest an error rate of 8%.

Now that the chairman is better informed about how pension plans operate, he is working with the General Accounting Office to conduct more detailed studies. If such studies indeed show that we are all at risk, we can all start worrying. Until then, the sky remains in place.

Population Seesaw

It is an unusual year when Social Security policy is not a subject of debate in the United States, and in recent years the attacks on the system have been particularly noisy. The paramount recurring theme is that a growing number of elderly recipients will become an intolerable burden on the working Americans who pay Social Security taxes. The system will be bankrupt early in the 21st century. There will be generational conflict, and social upheaval.

This gloom-and-doom scenario is attributed to the future projection that a "modest number" of working Americans will have to support a "huge older population" drawing Social Security benefits. Feeding into this concern are reports such as the one published by the U.S. Bureau of the Census which noted in a release entitled "Nation to Reach Zero Population Growth by 2050," that the percentage of people 65 years old and over will increase from 11.4% in 1981 to 21.7% by year 2050 (9). According to its authors, an important finding of their analysis is that the ratio of the working age population (defined as 18 to 64 years) to the retirement age population (defined as 65 years and over) will decline from 5.4 workers in 1981 to just 2.7 workers by year 2050. And many other commentators and public figures have repeated, in one form or another, these same ominous figures.

But this gloomy scenario occurs because of a lack of knowledge of the demographic effect of age shifts in the population. None of these prophets of doom have analyzed the total dynamics of a changing age distribution and how an *entire* population is supported. This is because they do not understand an essential element: that age distributions often resemble a seesaw with a fulcrum that is the "working" population and with each end of the seesaw balanced, respectively, by the elderly and the young. How does this seesaw work?

Those who are in full cry about the (possibly) unsustainable costs of Social Security claim that when the number of workers *per retiree* falls to about 2.7 workers in year 2050, the system will be bankrupt. But when the critics compare only one end of the seesaw—the number of dependents aged 65 and over—with the number of workers per retiree, and forget the other end—the worker's *children* who must be supported—they commit a misuse of statistics which has serious consequences.

Children are not self-supporting, and it has been the custom for most (not all!) of the support of dependents under 20 to be paid by the working-age class as parents. Thus, the working population, which is mostly aged 20 to 64 years, supports this younger population *as well as* the older population of Social Security beneficiaries. And remember, most of the contributions to the support of the dependents age 65 and older have been paid as Social Security taxes contributed by *both* those of current workers and the ultimate recipients when they were in the work force.

It is the total burden of support that is the issue for the working population, and the number of the elderly is only one of the components of this burden. For example, throughout the 19th century, about 50% of the total population was aged 20 to 64. This age group, the prime working force, supported the *total* dependent part of the population: the children and the elderly, about 50%. It can also be shown that from 1880 to 1990, while the proportion that was older (and no longer working) *increased*, the proportion of younger dependents *decreased*, still keeping the seesaw in balance.

Thus, the issue is the *total* number of dependents each worker will have to support. Since the national will is to support dependents, the political question is: By what mechanism will the funds for their support be distributed? As the relative proportion shifts, the mechanism of the funds transfers, and the allocation of those funds may be changed. This phenomenon has been observed by others as far back as 1978. More recently, in a 1997 letter published in *Science*, Bernard Brown noted that "Both child and elderly population levels affect the economy and the federal budget...[and] should be considered in the current debates over Social Security and Medicare" (10).

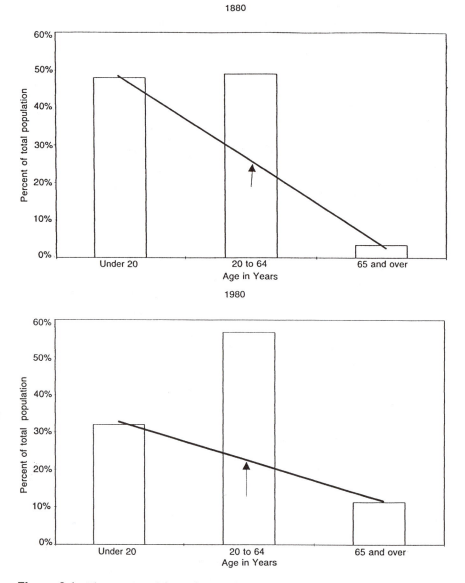

Figure 3.1   The seesaw of dependency. The dependent proportion of the population stays approximately constant, but the proportion of dependents is shifting from lower to higher ages as shown by the balance of the dotted seesaw line.

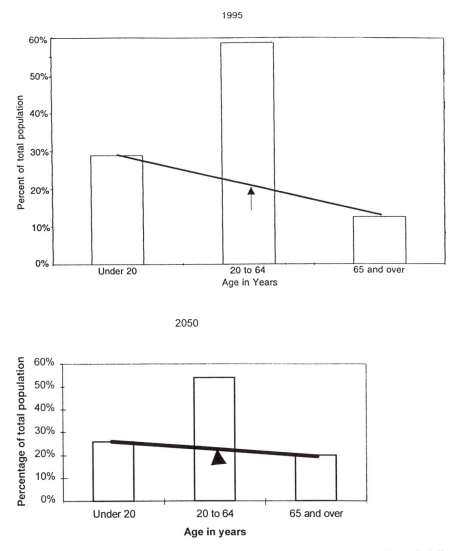

The birth rate of the United States is declining and the number of children is projected to increase by about 30 million from 1990 to year 2050. At the same time, the number of retirees is projected to rise, at a higher rate, by about 48 million; thus, the total number of dependents, young and old, will rise by 78 million. But not to worry: the forecasters say that the total number of workers is projected to increase by 70 million!

Figure 3.1 shows that the seesaw is working its magic: about 50% of the population will continue to support the other 50%, and the dependency bur-

Table 3.1 The Seesaw at Work. Number of Workers (Aged 20 to 64) per Person over 64 and Number of Workers (Aged 20 to 64) per Dependent Person Under 20 and over 64 by Year, 1880–2050

Year	Number of workers (aged 20 to 64) per person over 64	Number of workers (aged 20 to 64) per dependent person under 20 *and* over 64
1880	14.4	0.95
1980	5.0	1.31
1995	4.7	1.25
2050 (projected)	2.7	1.18

den on the work force will not dramatically increase. This relationship has held for over 100 years. In Table 3.1 you can see the consequences of the seesaw. Although the number of workers per person over 65 declines from 14.4 in 1880 to a projected low of 2.7 in 2050, the number of workers per dependent (persons under 20 and over 64) rises in the hundred years from 1880 to 1980, and then declines slowly from 1.31 in 1980, to 1.25 in 1995, and is projected to fall to 1.18 in 2050. The decline in the number of workers per dependent from 1995 to 2050 is thus projected to be less than 5%. It should be possible to deal with this slight decline by less than draconian measures (11).

We are left with some troubling questions. Why do politicians and some experts misinterpret the forecasts about the changing age distribution of the United States in a way that makes the increase of older dependents appear to be the cause of an "intolerable" burden on working Americans who pay for Social Security? In the absence of proof of a deliberate attempt to advance a political position, it might be no more than lack of knowledge of the subject matter. We suggest that our pundits take a trip to the nearest playground.

SUMMARY

Numbers do not interpret themselves. A number is a number—no more and no less. A number makes sense and can be understood only in light of some particular subject matter. The writer or analyst must know the subject matter before the numbers can be explained.

W. Edwards Deming, the famous American statistician who revolutionized Japan's system of quality control,* said it this way:

---

*Japan's highest award for contributions to the field of quality assurance is called the Deming Award.

Statistical practice is a collaborative venture between statistician and experts in subject matter. Experts who work with a statistician need to understand the principles of professional statistical practices as much as the statistician needs them.

Sometimes, as you will see in the examples we give in this book, the statistician *is* the subject matter expert. This is no problem when the statistician is really the expert, as often happens in demographics and social science. But when the statistician *thinks* that he or she is sufficiently expert and is not, misuse is always waiting in the wings.

4
Definitions

INTRODUCTION

Crime nose-dives. Poverty soars. Teenage pregnancy declines. Unemployment goes down. Childhood asthma is an epidemic. Executive compensation skyrockets. The number of driving-while-drunk convictions increases. Prevalence of child abuse remains constant. These are among the many statements about crime, health, poverty, employment, and the myriad other social phenomena which affect our daily lives. But these statements are without substantive meaning unless supported by statistics.

What is that substantive meaning? Any statistic we compute is an artifact, produced by human thought and computation. To know the meaning of a statistic for certain, you must find out how a general term (i.e., "crime," "poverty," "teenage pregnancy," "executive compensation," "unemployment") is defined. No matter how much effort we invest in collecting, refining, or improving the data, the resulting conclusion is based on a statistical definition, and that definition must be explicitly stated.

For example, what does "crime" mean? What is the definition of the measurement used to measure "crime" which is "nose-diving"? Is it the decrease in the offenses reported to the police obtained from police records? Or is it the decrease in victimization rates, obtained by surveys that seek out victims? How much credence do we give to either measurement? What is an "offense reported" and what is a "reported victimization"? How do we define these terms? We must dig deeper for an answer. Here's how the U.S. Census Bureau defines "crime."

There are two major approaches taken in determining the extent of crime. One perspective is provided by the FBI through its Uniform Crime Reporting Program (UCR). The FBI receives monthly and annual reports from law enforcement agencies throughout the country, currently representing 98 percent of the national population. Each month, city police, sheriffs, and State police file reports on the number of index offenses [murder and nonnegligent manslaughter, forcible rape, robbery, aggravated assault, burglary, larceny, motor vehicle theft, and arson, which was added in 1979] that become known to them. . . .

The monthly Uniform Crime Reports also contain data on crimes cleared by arrest. . . .

National Crime Victimization Survey (NCVS)—A second perspective on crime is provided by this survey . . . of the Bureau of Justice Statistics. Details about the crimes come directly from the victims. No attempt is made to validate the information against police records or any other source. . . . The NCVS measures rape, robbery, assault, household and personal larceny, burglary, and motor vehicle theft. The NCVS includes offenses reported to the police, as well as those not reported. . . .

Murder and kidnapping are not covered. . . . The so-called victimless crimes, such as drunkenness, drug abuse, and prostitution are also excluded, as are crimes for which it is difficult to identify knowledgeable respondents or to locate data records [1, pp. 197, 198].

These two sources of statistics—the Uniform Crime Survey and the National Crime Victimization Survey—may differ, as you can easily see from the definitions. Which statistic do you use to determine if "crime" is soaring? It's impossible to answer this question out of context. When you know the purpose of your analysis, you can choose the most useful statistic for that purpose. You may even decide to use several statistics, taking into account the several definitions.

PRELIMINARIES

Who's Who?

Do you define a "manager" by job title? If so, you include persons with responsibility for a part of a small office along with persons with responsibility for billions of dollars in projects. Many job titles given for similar occupations do not necessarily match salaries, responsibilities, and authority. But more importantly, in many job situations, a person's title is based only upon self-description.

Most of our information on occupations comes from surveys such as the Census and the Current Population Survey, in which the respondent is asked to name his or her occupation. Sometimes there is no problem: A bookkeeper is a bookkeeper. But what about the head of the accounting department which includes the bookkeeper? That person can report "accountant" or "manager." The self-reported title may or may not agree with the personnel records of the company. These self-reported designations are then classified into whatever formal system is in use at the time. In addition, the classification system is often changed by the surveyors to fit the current practices. Since the apparent needs of employment are always changing (we no longer list carriage or harness makers as occupational categories), the occupational classification system is continually reexamined and revised as necessary. As noted by the U.S. Bureau of the Census:

> Comparability with earlier census data — Comparability of industry and occupation is affected by a number of factors, a major one being the systems used to classify the questionnaire responses. . . . The basic structures were generally the same from 1940 to 1970, but changes in the individual categories limited comparability from one census to another. These changes resulted from the need to recognize the "birth" of new occupations, and the "death" of others. . . . In the 1980 Census . . . the occupation classification . . . was substantially revised [2, Appendix B, p. 8]. (The Census Bureau has changed some job classifications, but the basic text is the same in 1997 as in 1980.)

"Definition" and "classification" are the twin names of the game. Watch both.

You Are What You Say You Are

Race is another concept for which there are no natural or correct definitions. We are not even sure what race means. If race means homogeneity of genetic structure, then we are indeed in deep definitional trouble. It is likely that every person alive is a hybrid in the genetic sense.* When we ask individuals to report their own perception of their race, which is the usual way to obtain data on race, we do not have their genetic code available.

In the United States and some other countries, skin color is the basis for classification of race, which is a subjective evaluation. Thus, many people listed themselves as "brown" in the 1980 census, a category that does not appear in the official census list. Perhaps the designation "white" is also sus-

---

*The announcement in February 1997 of the birth of a cloned Scottish sheep, Dolly, holds the possibility that humans could be cloned someday. If this does indeed occur, our statement might no longer be valid. We promise to revise our statement accordingly in a new edition when the time comes.

pect: As Mark Twain pointed out some time ago, most so-called white people are really pink!

Skin color is also a social definition of race, as you can see from the history of race definition in America. The U.S. Constitution originally specified that congressional representatives were to be allocated according to a decennial census, in which the categories were defined as follows:

> Representatives . . . shall be apportioned among the several States . . . according to their respective numbers, which shall be determined by adding to the whole number of free persons, including those bound to service for a term of years, and excluding Indians not taxed, three fifths of all other persons [3].

Since the slaves were "black," the Constitution defined three "race" categories for census purposes: free persons (which could include a small number of blacks and all indentured servants, and those few taxed Indians); untaxed Indians; and "other persons" (who were mostly black slaves). As the nation grew and changed its social structure, those social changes affected the census categories and concepts of race. In the 14th Amendment to the Constitution, the categories for apportionment were reduced to two: "counting the whole number of persons in each State, excluding Indians not taxed."

By the middle of the 19th century the decennial census had become a source of general information; the racial classification of white and black was retained. By 1860 enough Chinese had immigrated to the United States to warrant adding the racial category of Chinese to that census. In 1870, following the immigration of Japanese persons, that category was added to the census racial classification system. Anyone who claimed not to be white, black, or American Indian was given a separate "race" designation. One hundred years later, the 1980 census schedule listed the following categories: "White, Black or Negro, Japanese, Chinese, Filipino, Korean, Vietnamese, Indian (American), Indian (Asian), Hawaiian, Guamanian, Samoan, Eskimo, Aleut, and Other." The respondents made their own choices from this list. Thus persons who view themselves as "Brown" could check "Other Race." Others are counted as "race, not elsewhere classified," and the 1980 census reports 5.8 million Americans chose this classification (2, Table 74).

In 1990, the categories were simplified to White, Black, American Indian or Aleut Native, Asian or Pacific Islander, and Other. But under the classification "Asian or Pacific Islander," for example, the Census Bureau has added under the subtitle "Asian": Chinese, Filipino, Japanese, Asian Indian, Korean, Vietnamese, Cambodian, Hmong, Laotian, and Thai. In 1990, there were 9.8 million Americans who chose the classification, "Other Race."

In July 1997, a federal task force "said 'no' to the idea of a multiracial US Census category" (4). The Task Force said that such a category would "create another population group ... and add to racial tension. . . ." The task force recommended that persons filling out the form could select more than one category, rather than having to choose whether to identify by either the mother's or the father's race (which might also be mixed). This type of identification, however, could create some problems involving, for example, how to determine the proportion of blacks and whites or Asians and blacks in a given geographical area (1, page 4).

The number of Hispanics in the United States increased greatly after World War II, and since there was no simple way to fold them into a racial classification, a separate question was added to the schedule to determine whether the respondents see themselves as Hispanic. This section included prelisted categories, i.e., Mexican, Puerto Rican, Cuban, and so forth. Published tables carry this category separately from those of the various races.

Defining "race" is an age-old problem which is still a thorn in the sides of census takers, immigration officials, and policy makers.

Has There Been a Change or Hasn't There?

In Prices?

Do we overstate the artifactual nature of definitions? When major industries can improve their bottom line by influencing definitions, it becomes clear just how much of an artifact a statistic is.

In 1980, a headline, "Producers' Prices Show Slight Drop, First in 4 1/2 Years," announced happy doings in the automobile industry (5). "The decline was centered on motor vehicles, food and energy," says the text. But the final paragraphs of the article reveal that there was a change in definition which caused the drop. If the definition had not been changed, the Index of Producers' Prices would have shown a rise twice as large as the decline noted in the headline.

What was it about the change in definition that caused the "slight drop"? It was changing the definition of automobile *price* to include the effect of model year-end discounts given on new automobile sales. In the year before the change in definition, the price of automobiles was entered into the index according to the "sticker," or list, price. With the exception of Saturn dealers, most new-car dealers negotiate a lower selling price based on several factors including the value of the trade-in car, the accessories purchased, and a discount. The discount rises at the end of the model year, when knowledgeable buyers take advantage of the dealers' anxiety to reduce inventory before the arrival of the new models. Thus, the apparent price of the car drops.

Although not applicable to all car sales, this rise in the discount affected the average reported price of automobiles. Here's how it worked. Suppose that the average sticker price for a given year is $20,000. If no account is taken of discounts in the computation of the index of producers' prices, the figure $20,000 is added to the prices of other items which comprise the index. However, if an average discount of 20% is offered in the last quarter of the year, then the average automobile price added to other prices in the index would be $19,000. This reduction is based on an average automobile price of $20,000 for the first three quarters and 20% less, or $16,000 for the last quarter. The amount entering into the producer's index now will not be the sticker price of $20,000 but instead, three-fourths of $20,000 plus one-fourth of $16,000, or $19,000. Thus, the automobile price entering the index would be 5% less than under the original definition, enough to give the impression of an apparent reduction in automobile producer prices.

This drop did not represent a change in producers' prices; it represented the change in the definition of "price." Thus we see that "correctness" is arbitrary in the statistical context.

Sixteen years later, in 1996, average car prices once again made headlines. In late 1994, the Commerce Department released its calculation that the average new car price was just over $20,000, and there was public concern that cars were becoming too expensive. Subsequently, "The [Commerce] department recalculated its figures with much greater precision this year [1996] after protests from automakers here, who were upset by the suggestion that they were pricing themselves out of the market" (6).

The "greater precision" involved a change in definition of average car price which brought the average price down to $18,000. The major definitional change was to add in the significantly lower car prices paid by "fleet purchasers." Unfortunately, although they may hear about lower average prices, individual purchasers will still pay a higher price than the fleet purchaser. The Commerce Department's spokesman "insisted that auto industry pressure played no role in the agency's eventual decision to choose a new method for its calculations [actually, a new definition] that happens to produce a result more pleasing to auto makers" (6).

Draw your own conclusion, and keep your eye on the definition and those who make it.

In the Number of Poor People?

We can correctly (or incorrectly) count the number of families below some arbitrary poverty line, but we cannot correctly determine the number of poor, for:

> Families and unrelated individuals are classified as being above or below the poverty level using the poverty index originated at the Social Security Administration in 1964 and revised by Federal Interagency Com-

mittees in 1969 and 1980. The poverty index is based solely on money income and does not reflect the fact that many low-income persons receive noncash benefits such as food stamps, Medicaid, and public housing. The poverty index is based on the Department of Agriculture's 1961 Economy Food Plan and reflects the different consumption requirements of families based on their size and compositions. The poverty thresholds are updated every year to reflect changes in the Consumer Price Index [1, p. 442].

How dependent is the number of the poor on the definition? Read this:

The following technical changes to the thresholds were made in 1981: (1) distinctions based on sex of householder have been eliminated; (2) separate thresholds for farm families have been dropped; (3) the matrix has been expanded to families of nine or more persons from the old cutoff of seven or more persons. These changes have been incorporated in the calculation of poverty data beginning with 1981 [1, p. 442].

These "technical changes" have considerable effect on the number of poor. Anyone wishing to compare the pre-1980 numbers of the poor with post-1980 numbers will be comparing apples with oranges unless they make corrections, which at best can only be estimates. It is no wonder there is so much misuse of poverty statistics in public debate.

Preventing misuse takes constant attention. It is possible to do better. The poverty measurement concepts we describe above are over three decades old. It is not surprising that the public debate about the poverty threshold is heated and continuous. Starting in 1992, with a charge from the Joint Economic Committee of Congress, and continuing with funding from several governmental agencies, a panel undertook a study "to address concepts, measurement methods, and information needs for a poverty measure" (7, p. xv). The panel's recommendations include:

The official U.S. measure of poverty should be revised to reflect more nearly the circumstances of the nation's families and changes in them over time. The revised measure should comprise a set of poverty thresholds and a definition of family resources—for comparison with the thresholds to determine who is in or out of poverty—that are consistent with each other and otherwise statistically defensible. . . . [The government] should institute a regular review, on a 10-year cycle, of all aspects of the poverty measure. . . . When changes to the measure are implemented on the basis of such a review, concurrent poverty statistics series should be run under both the old and new measures to facilitate the transition [7, pp. 4–5].

We believe that if the panel's recommendations are followed, we can expect a much better use of statistics in the public policy debates about poverty.

When Is a Tax Audit Not a Tax Audit?

When it's a "notice." According to a 1997 article in *The New York Times*, it would appear that taxpayers in poor districts, such as Mississippi, are more subject to audit by the I.R.S. than those in wealthier districts, such as New York (8). The figures released by the I.R.S. did indeed sound as if tax audits for those who earn more than $100,000 per year have dropped dramatically by one-third while people who earn less than $25,000 have seen a dramatic increase of nearly one-half in audited returns. Despite the fact that audits of large companies "produce 58 percent" of revenues generated by these audits, the number of audits has dropped (the I.R.S. cried poverty, not prejudice, when asked about this figure, claiming that their budget had been cut). Are poor people being targeted? Not at all, says the I.R.S., even as it admits that the numbers are accurate. It seems that the I.R.S. has become more diligent in sending out notices to those people whose taxes have been withheld but who have not sent in their returns. So how do we define "audit"? If it now includes "notices" or "reminders" sent to those who have not filed returns as well as the usual requests made to those suspected of malfeasance or ignorance, how will this affect the statistics on the number of "audited" returns?

NUMBERS AND CONCLUSIONS IN CONFLICT

Is the stock market rising, falling, or static? When we ask this question we are trying to determine the general movement of stock prices: whether the market is a bull, bear, or sloth market. To answer this question, we form indexes by combining selected stock prices. These indexes are usually in agreement about market movement. This congruency reinforces our faith in the value of the indexes. But what would we think of these indexes if the Dow-Jones, Standard & Poor's, and American Exchange indexes were in conflict about the market movements day after day?

If these statistics were in continual conflict, then we would cease to accept their conclusions and would demand their revision. Unless we were convinced that one was superior to all the others, we would draw no conclusions about market movement.

Let us look at some of the cases in which we have conflict among statistics that are supposed to measure the same entities, conflict so serious that we must question our ability to use these statistics in the formulation of policy.

A Case of International Discord

Comparisons of the United States with other countries are often made with respect to such vital statistics as the birth rate, infant mortality rate, income, unemployment, and so forth. But all countries do not use the same definitions or methods in collecting data. Thus, such comparisons are dangerous and full of traps for the unwary. We cannot review the procedures of all countries, but can only— as we do elsewhere—give an example to show you just how dangerous it is to accept these comparisons without question. A word to the wise: When you hear of, or make, such comparisons, we advise you to study carefully the technical appendixes and introductions to the books containing the published information.*

The introduction to the *United Nations Statistical Yearbook* has the following warning:

> One of the major aims of the *Statistical Yearbook* is to present country series which are as nearly comparable as the available statistics permit. . . . There are certain limitations . . . of which the reader should be aware. For example, a direct comparison between the official national product data of countries with market and centrally planned economies would be misleading owing to differences in the underlying concepts [9, p. xviii].

When the above statement is translated into straight talk, it says that countries with market economies (such as the United States) and those with planned economies (such as the People's Republic of China) define economic quantities differently. While you might be able to make adjustments to improve comparability, you still must make comparisons cautiously, *very* cautiously, to avoid misusing statistics.

Another example of noncomparable definitions comes from a different United Nations publication:

> Most of the vital statistics data (births, deaths, marriages, and divorces) published in . . . [the *United Nations Demographic*] . . . *Yearbook* come from national civil registration systems. The completeness and accuracy of the data which these systems produce vary from one country to another . . . [9, p. 13]. A basic problem facing international comparability of vital statistics is deviation from standard definitions of vital events. An example of this can be seen in the cases of live birth and foetal deaths. In some countries or areas, an infant must survive for at least 24 hours before it can be inscribed in the live-birth register. Infants who die before the expiration of the 24-hour period are classified as late foetal deaths and, barring special tabulation procedures, they would be counted either as live births or as deaths. But, in several other countries or areas, those infants who are born alive but who die before registration of their birth (no matter how long after birth registration may occur) are also considered as late foetal deaths [9, p. 15].

---

*Herbert Jacob has written an excellent, if somewhat technical, guide for such evaluations. See Herbert Jacob, Using published data: Errors and remedies, *Sage University Paper Series on Quantitative Applications in the Social Sciences, 07-001.* Sage Publications, Beverly Hills and London, 1984.

The following simple definition of a live birth is internationally agreed to and defines the birth which is used in computation of the birth rate: If the new-born shows any evidence of life it is counted as a live birth even if it lives for only a few moments. Whenever a person dies—whether at the age of one day or less, or at the age of 100—it is registered as a death. Since some countries deviate from this definition of a live birth for statistical purposes, you should make international comparisons of birth-related data only when you know the precise definition used by the countries you are comparing.

It is not only for demographic data collection that different nations use different definitions for ostensibly the same statistic. Examples in manufacturing, immigration, medical administration, unemployment, and so forth, can be given. But the careful reader is rarely at a disadvantage, for national and international statistical compilations usually give definitions and cautions at length. In cases where definitions are not given explicitly, it is often possible to track them down. Problems arise when the headline writer, the Dr. Nowall, or the politician, in an effort to make a point, omits, accidentally or deliberately, cautions about changes in definition from their statements.

The First Casualty

To paraphrase a well-known witticism, in times of election, statistics are the first casualty. In the hard-fought 1996 election campaign for a U.S. Senate seat in New Jersey, candidate Zimmer accused candidate Torricelli, a congressman, of having voted for "the largest tax increase in history." Although this may appear to be no more than election rhetoric, as ephemeral as a campaign promise, it is a statistical statement. How are we to define the size of a "tax increase"? The definition can have a great effect on the resulting statistics.

The bill Zimmer was referring to was President Clinton's five-year economic plan (1993) which amounted to $241 billion, an all-time numerical high (10). You can simply take the "bottom line" total of the congressional bill which levies the increased tax, and by this definition, Congressman Torricelli did vote for the largest tax increase in U.S. history.

Does this definition get at the *value* of the tax increase? Mr. Zimmer is appealing to the judgment of the voter, whose concern is the cost of living. Inflation reduces the purchasing power of the dollar. One common way to correct the value of the U.S. dollar is to *deflate* the current amount for inflation, using the Consumer Price Index (CPI).* Thus, taking the CPI into ac-

---

*We take the CPI "into account" by expressing value of money in terms of its value to the consumer in purchasing goods. The CPI in 1982 was 96.5; in 1993, it was 144.5. Thus, $1.00 in 1993 bought only two-thirds as much as $1.00 in 1982 (96.5/144.5=0.67). It doesn't pay to keep dollars under your mattress.

count, President Reagan's 1982 bill, in 1993 dollars, came to $322 billion, a larger tax increase.*

We have implicitly defined the size of a tax increase in terms of its current value to a voter. There are other possible definitions for this measure. We might compare it to the Gross Domestic Product (a measure of national economic activity). Or, we might define "largest" in terms of the per-capita inflation-adjusted value. We think that sound political decisions (including voting) are best made in cases like this by reviewing the several possible definitions, and also by questioning the candidate to find out what definition the candidate is using. That's not an easy task in the heat of an election, but the intrepid voter or Letter-to-the-Editor writer can make the point and, hopefully, flush out the candidate.

When Is a Capital Not a Capital?

As the preceding example illustrates, public policy debate is often expressed in terms of extremes. "Teen-age pregnancy is at an *all time high*," "Crime rate *lowest in 25 years*," "Drug use is *higher than ever*"—these are the cries. When advertisements for commercial products use comparisons, the ads often make statements such as "More pain relief than ever," and the statistician's gem, "Forty percent larger." You need a good sense of definition to evaluate any of these statements.

At his swearing-in ceremony, the new District of Columbia Police Chief, Fred Thomas, said that the District was getting a bad rap from "news reports" which created an image of the District as the homicide capital of the U.S. (11). In his statistical analysis, the chief argued that another community had a higher *per-capita* homicide rate. This is tantamount to saying that the size of the per-capita rate is how a "capital" is defined for homicide, and presumably other measures of civic welfare. He then went on to argue that other cities had a higher *total count* of homicides, which set up another definition, this one more favorable to the District. This definition presumably put these other communities in line for the designation "homicide capital of the U.S." He also argued that a better definition would take into account the high daytime population in the District, which does not appear in the population counts that form the basis of per-capita measurements. In short, if you don't get the result you like with one definition, try another.

---

*We have our own quarrel with the use of the CPI in this way. It is not a cost of living index, but a price index, based on a standard "market basket" of commodities. But the CPI will serve where the differences are large.

All of this shows the essential absurdity of such comparisons. If your purposes are to understand the social environment in a community, to judge its social services, and to form public policy through debate, we feel that such comparisons are of no real value, even when there is a single explicit definition for the comparative measure. If those are your purposes, the community is best served by studying many measures (of the same underlying process), all of which call for explicit and sensible definitions.

TB or Not TB?

A time series is a set of data taken at different times during some period. Typical time series are the shipments of automobiles, net trade balances, population, corporate profits, time deposits, and insect infestation. All of these may be reported by day, week, month, quarter, year, or decade. We are concerned with the problem of comparability that arises because of changes in definition. Take, for example, the following case, which was reported in *The Washington Post:* "The tuberculosis rate in the District [of Columbia] . . . dropped 30 percent last year to the lowest rate in 30 years, Mayor Marion Barry announced yesterday. . . . The number of reported cases in the city dropped from 341 in 1980 to 239 in 1981" (12). It is quite possible, as the director of the District's TB Control Program said, that the decline was due to the agreement in 1981 of physicians and clinics to use a new double-drug therapy. However, our evaluation of whether there was such an effect is clouded by the change in definition of a "reported" TB case. In 1980, "all suspected tuberculosis cases reported to the TB control clinic by doctors and health clinics were included in statistical reports unless tests specifically rule out the disease. . . . But in 1981, only medically confirmed cases were included."

It is not a misuse of statistics to change the definition of a statistic from one period of time to the next. In fact, there are cases where it would be a misuse not to do so! However, it *is* a misuse to fail to recognize, or state, that a change in definition might be one cause for a reported change in some important value.

Separating the Divorced

"Family values" have been a major issue in U.S. political debates in the 1990s. The divorce rate is one measure of the stability of the family, but how do we know if the "divorce rate" is rising or falling? More importantly, *which* divorce rate actually measures the concept of stability of marriage? This debate often becomes a squabble about apples and oranges because parties to the debate use different definitions of the divorce rate. To add some light and take away some heat from these discussions, we show two ways to look at marital stability. Each method corresponds to a unique definition which measures a particular statistical artifact.

You can easily find out how many marriages and divorces occur in a given year. The compilation of these basic vital statistics started in 1887 and the National Office of Vital Statistics made its first estimates for the period 1867–1876.

When a misuser, attempting to show how the divorce rate has affected family life, divides the number of divorces for a given year by the number of marriages for that year, we know nothing about the status of families. Marriages and divorces in a given year are only *events*. This statistic, which unfortunately is sometimes called "the divorce rate," is of value only to someone concerned with the relative frequency of these two kinds of events. It tells us nothing about the stability of marriages. The number might be of interest to marketers of certain products, but it does not illuminate debates about the family.

Social scientists, policy markers, and decision makers could well be concerned with the rate at which the population "generates" divorces. This rate is the number of divorces in a given year divided by the total female population of marriageable age (15 or older) in the given year. This is the logical basis for the rate because women under 15 are usually not candidates for divorce.

Figure 4.1 shows the value of this statistic from 1920 to 1994 (13, Table 146; 14, Table 1). From this graph, you can see that the divorce rate has been moving upward since the late 1930s. You can also see the pronounced peak at the end of World War II, and a period of relatively little change afterward until the early 1960s. It is clear that the divorce rate started its climb in 1963, moving upward at a near-constant rate until its peak of 22.6 in 1980. This recent value of the divorce rate is about three times the average rate prior to 1936 and slightly more than double the rate between 1950 and 1964. However, from 1980 to 1994, the rate has declined about 4%.

This definition of divorce rate may not be appropriate if the concern is for the stability of the legally sanctioned family. In this case, you might want to know the average duration of marriages before divorce or annulment. We show these statistics in Figure 4.2 as obtained from the *Statistical Abstract* (1, Table 150).

The next time you hear a speech about divorce rate, look at Figures 4.1 and 4.2, or their current equivalents. Then draw your own conclusions.

Something Is Fishy Here

As we have seen, it's one thing to not have a clear or explicit definition. It's another thing to ignore a definition. In February 1997, the Health Department

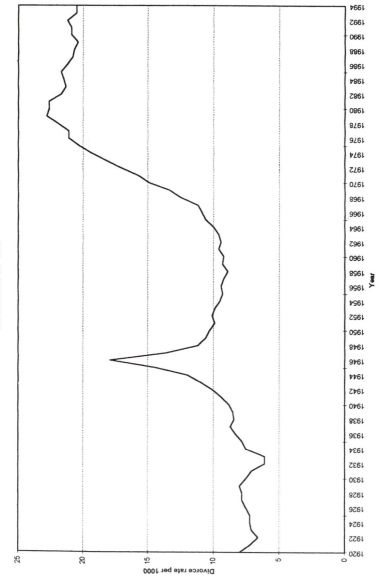

Figure 4-1 Divorce rate.

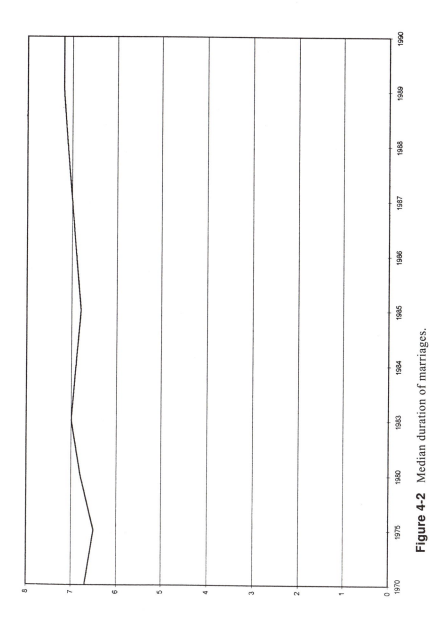

Figure 4-2 Median duration of marriages.

of Greenwich, CT, issued a caution to people concerning high levels of mercury in fish taken from the local Mianus River.

The official definition of mercury level requires making a comparison based on the results of testing *fresh* fish tissue. Instead, the technician used *dried* fish tissue, which gave a result that was five times higher than the level of mercury that the correct test would have shown.

Town officials, ever resourceful, claimed to have suspected "that the report was inaccurate, [but] they issued a notice reminding the public to adhere to the Department of Public Health's general advisory on eating fish" (14).

IS IT HARD TO BE INTELLIGENT ABOUT IQ TEST SCORES?

In Chapter 1, we discussed *The Bell Curve,* the 1990s best-seller on the relationship of race and IQ test scores. The authors' definition of intelligence is a single number, the score on the standardized Stanford-Binet test. This measure, called the IQ score, like any other test result, is a statistical artifact.

For support, the *Bell Curve's* authors reference a study of intelligence differentials between blacks and whites in South Africa. They say that the study showed that black African IQ scores are deficient. However, this study did not use the standardized IQ test; it used a different definition of IQ! The cultural bias in this artifact was considerable. The study's author reported several years before the publication of *The Bell Curve* that "the knowledge of English of the majority of the black testees was so poor that certain tests proved to be virtually unusable" (15, p. 100). *The Bell Curve's* authors were undeterred by this fact.

The Stanford-Binet test is a moving target, as are the SATs, LSATs, GMATSs, and other such tests. Its questions reflect the backgrounds of those who create the questions and how frequently those questions are changed. As Douglas A. Samuelson says in his insightful overview of intelligence testing:

> For example, tests which included such "general knowledge" items as
> the identity of New York's baseball team were used in the 1920s to sup-
> port the conclusion that recent immigrants from Eastern and Southern
> Europe were "inferior" to Americans of Western European ancestry [16].

Cultural bias is always a problem. It is usually thought that cultural bias is a complaint made by underperforming minorities. However, one of us heard a member of a usually high-performing group complaining about a question that involved knowing that the acorn was the seed of an oak tree. This individual argued that he could recognize an acorn on sight, but having spent most of his childhood in an urban area, he never realized, as would any suburban or rural resident, that it was the seed of the oak tree. Cultural bias can cut in many directions, and its effects show the artifactual nature of statistical measures in social science.

SUMMARY

For the Reader

Take nothing for granted. Find and understand the definition of key statistics given in the article or report.

If no definition is given, don't give credence to the results reported until you get one. If the issue is important to you, contact the author and get a definition.

If you can't understand the definition, seek an explanation, either from the author or from other sources.

If the definition you get doesn't make sense to you, don't accept the results.

If two or more statistics are compared, check the definitions for comparability. Be persistent in seeking out differences and the methods of adjustment.

For the Author

Give definitions in your reports and articles. And when comparing two statistics, give full information about differences in definitions and the nature of adjustments.

5
Quality of Basic Data

Her taste exact
For faultless fact
Amounts to a disease
    —W.S. Gilbert

Measurement does not necessarily mean progress. Failing the possibility of
measuring that which you desire, the lust for measurement may, for example,
merely result in your measuring something else—and perhaps forgetting the
difference—or in ignoring some things because they cannot be measured.
    —George Yule

INTRODUCTION

"Garbage in, garbage out" is a sound warning for those in the computer field;
it is every bit as sound in the use of statistics. Even if the "garbage" which
comes out leads to a correct conclusion, this conclusion is still tainted, as it
cannot be supported by logical reasoning. Therefore, it is a misuse of statistics. But obtaining a correct conclusion from faulty data is the exception, not
the rule. Bad basic data (the "garbage in") almost always leads to incorrect
conclusions (the "garbage out"). Unfortunately, incorrect conclusions can lead
to bad policy or harmful actions.

   It is hard to get good basic data, even in simple, straightforward situations and with the best of intentions. National census enumerators in developed countries may not be able to reach all inhabitants, or may accept inadequate results in order to meet a production quota. Data entry operators reading
data from documents may incorrectly enter it. Quality control inspectors may
misread meters and gauges. Survey respondents may not understand the questions or may inadvertently or deliberately give biased responses. Even counting objects is subject to human error.

Consider the following mythical example, which is typical of a situation that often arises in practice.

Not So Fast

Kenneth F. Capon, Dr. Nowall's uncle, runs a fast-food store. When the store began experiencing problems, Capon hired a business consultant who suggested that Capon collect data on the flow of patrons as basic information for a plan for future operations. "No need to collect data," says Capon; "I've been running this shop for several years. I don't have to spend time counting customers to know that I get about 20,000 customers a week. Divided by 7, the number of days a week the store is open, this comes to 2857 per day."

When Capon says that he gets an average of 2857 customers per day, the very appearance of the figure of 2857 per day implies counting for a precise value. It's not 2856, not 2858, but exactly 2857, and it might well be accepted as a precise value and used as such in future calculations and analysis. However, this value has its origin in a guess: it is a "mythical number," and, as Max Singer has pointed out (1), mythical numbers can have an amazing vitality, persisting in the face of all future evidence, as we show in this chapter and in Chapter 13.

Capon's consultant is both dubious and cautious, and he persuades Capon to count the patrons. As the owner, Capon doesn't want to stand at the door and do nothing but count patrons for a week, so he hires a high school student. The student counts conscientiously the first day, but soon becomes bored and is frequently distracted by conversation and the passing scene. When distracted, the student either records none of the entering patrons or guesses a number for the periods when he is not actually counting. The result is an inaccurate number which is simply wrong. But how wrong? Too high or too low, and by how much? Five percent may not matter for some purposes, but for ordering or planning future purchases, a 5% error may have serious economic consequences for Mr. Capon. Thus, even as simple a task as counting customers can be botched by poor management or incomplete instructions. Whether the student enumerator was inattentive or poorly instructed, bad data could have been collected; any analysis of such data was suspect.

Add to this the other complexities of simply counting patrons. Does Mr. Capon want to know how many people came into the store, or how many people made purchases? Should he instruct the young student to count family groups as one or as individuals?

In addition, Mr. Capon seems to have a rather hazy idea of what his goals are. Just counting patrons may not be enough. He and his consultant need to sit down and make a sensible business plan and then, based on still more data such as neighborhood demography, competition, and so forth, make some decisions. Such decisions again go back to basic data, such as neighborhood demography, which may be inherently flawed. Depending on the neighborhood, even official census data may be incorrect. Both Capon and his consultant must exercise both caution and skepticism before accepting these data as well.

BAD MEASUREMENT AT ALL LEVELS

A Cockamamie Universe?

For some time, astronomers have been puzzled by the paradox that the universe appeared to be younger than its oldest stars. The stars appeared to be 14 billion years old and the universe in which they resided, a mere 9 billion. This is indeed a bizarre result, so it is reasonable to ask questions about the measurement of stellar ages. The basic yardstick for astronomical star measurements is a comparison of stars called Cepheids, whose brightness is directly related to their pulses. Astronomers find two Cepheids with the same pulse and use the brightness difference to estimate their physical separation.

In 1997, astronomers reported on measurements of the positions of over a million stars, to varying degrees of accuracy and precision, made by Hipparcos, a European satellite. Free of the effects of the atmosphere and gravity field of the Earth, these measurements are better than any Earth-based measurements. The measurements made by Hipparcos indicate that the Cepheids were about 10% farther away than prior estimates of their distance. As a result, Michael Feast, of the University of Cape Town, can assert: "I hope we've cured a nonsensical contradiction that was a headache for cosmologists. . . . We judge the universe to be a little bigger and therefore a little older, by about a billion years" (2). Professor of Astronomy Robert Kirschner at Harvard University also sees hope: "We are finally getting to the point where we don't disagree on the distance scale by a factor of two, but by 10%. That's good. It means science is growing up" (2).

So how did the apparent paradox, that troubling contradiction between the ages of the stars and their universe, arise? "Until Hipparcos, the cosmic distance scale rested on *well-informed guesses* [our emphasis], said Michael Perryman, project scientist for the European Space Agency" (2).

The issue, of course, is not that since we now have better technology for measuring the position of the stars we feel free to criticize the original figures, but that the previous measurements were *well-informed guesses*. Admittedly in hindsight, we feel that the contradiction of about 4 billion years in the ages of universe and the stars was enough to indicate how poor the data were, not that the universe was perverse.

Beware the "well-informed" guess!

. . . While Here on Earth

In the spring of 1997, the current administration of the City of New York announced that 83% of the city's streets were "acceptably clean," the highest rating in 23 years. Not surprisingly, a mayoral election was due in the fall of the same year. This heartening statistic was contradicted by a *New York Times* report that 36% of the city's street were at least "acceptably clean."

How do we explain this large difference? It could be differences in definition, but the public reports of the resulting debate do not mention such a disagreement. The difference is attributed by both groups to the method of data collection and a weighting process used in the analysis. We do not deal with the weighting process as a cause of this difference, but with the lesson to be taken from the difference in data collection methods:

> William P. Canevale, director of the Mayor's Office of Operations, who oversees the unit that rates streets, defended the city survey. . . . [He] said that there were several reasons why [the results] differed so markedly. . . . One reason, he said, was that the city's raters were more experienced, more objective and better trained than the newspaper's reporters [3].

The *Times* survey was a sampling made by two reporters who, using a rating devised by the Urban Institute in Washington, DC, surveyed 800 randomly chosen streets, walking the streets and rating them. The city's team rated the streets from a moving car. Can you compare data on the cleanliness of streets obtained from a moving car to that obtained by pedestrian examiners?

DATA WITHOUT SOURCES

How Many Illegal Aliens?

When data are reported without a source, you can't take them seriously until you find a confirming source. Again and again, we find attention-getting statements that test credibility, and are given without sources. There is no shortage of examples of such misuses of statistics in headlines. For example, the headline of a staff-written article appearing in the *Denver Post* in 1980 says, "20 Million Illegal Aliens Get Reprieve." The article goes on to say, "Most of the estimated 20 million illegal aliens in the United States have been granted a reprieve from arrest, prosecution and deportation through a national policy decision made by the U.S. Census Bureau" (4).

Counting illegal aliens is not easy, and the purpose of the Census Bureau's policy was to improve the estimate of the number of illegal aliens for counting purposes by reducing their resistance to identifying themselves. David Crosland, acting Immigration and Naturalization Service commissioner, said the basis for the policy was "to insure that census operations take place in an atmosphere conducive to complete participation and disclosure of information by all groups" (4).

So what of the claim in the *Denver Post* article that there were 20 million illegal aliens living in the United States in 1980? Is it possible that about 9% of the U.S. population comprised illegal aliens? To some, this may have sounded like a remarkably high proportion of the U.S. population, but perhaps even this number is low. Or could it have been a reasonable estimate? Unfortunately, the article provides no source for this number. Thus we don't know who supplied this number or how it was obtained. How can we give it any credence?

Jeffrey S. Passel, then chief of the Population Analysis Staff of the Population Division of the U.S. Census Bureau, gave us a credible estimate in a paper he presented at the 1985 Annual Meeting of the American Statistical Association. He found that:

> Analytic studies of the size of the undocumented population indicate that it is significantly smaller than conjectural estimates [such as those of the *Denver Post* article] suggest. The studies reviewed here and the synthesis involving new data point very strongly to a range of 2.5 to 3.5 million undocumented aliens as the best figure for 1980 [5, p. 18].

So much for the "20 Million Illegal Aliens."

How Many Deaths from Artificial Food Additives?

The director of the Center for Science in the Public Interest, Michael Jacobson, was reported by the *Philadelphia Bulletin* to have said: "I'd estimate that a maximum of 10,000 to 20,000 deaths per year could be attributed to artificial food additives" (6).

Dr. Melvin Bernarde, of the Department of Community Medicine and Environmental Health at the Hahnemann Medical College and Hospital of Philadelphia, felt that if that many people died each year due to cancer-causing food additives, he would call it an epidemic and said:

> Few are going to question where or how Jacobson obtained his figure of 10,000 to 20,000 deaths per year. It will be accepted as fact—because it's in print. I've tried to corroborate his figures but can't. Not because no one will give me the data, but because no one appears to have them. Yet they are, thanks to Jacobson, now part of the public record—to be quoted and requoted [7].

Once again, the attention-getting number may be high, low, or even fairly accurate. But without a source, we can't make our own evaluation and the number is worthless for making a decision about personal or public policy in regard to food additives.

How Long Do You Think You Will Live?

In a laudatory article about Dr. Michael Colgan, a nutritionist from New Zealand who has developed an "individualized vitamin-and-mineral program," we find the following statement:

> The urgency of Colgan's mission—and it can be called that—is most clearly understood against the backdrop of nutritional statistics that dwarf our attempts to "eat right and stay healthy." Here's a small sampling:
>
> In both the United States and the United Kingdom the average life expectancy of an adult, twenty-five, has not changed for more than 30 years [8].

There are no U.S. or U.K. government sources for this statement given in the article. However, we can easily check this statement for the United States. The official figures show that the average number of years of life lived beyond age 25 ("life expectancy at age 25") for the total U.S. population increased from 46.6 years in 1949/1951 to 50.7 in 1980, an increase of 4.1 years. Life expectancy at age 25 increased for both men and women and for both whites and nonwhites (9, p. 8; 10, p. 74). The author of the article not only gives no source, but the figures in the statement quoted above are wrong. In view of this, how much credence can you give to the other statements and nutritional recommendations in the article?

We encourage skepticism, but we do not want to encourage unfairness. We don't know what Dr. Colgan and the nutritionists quoted in the article actually *said*; we only know what the author of the article *reported*. Thus, without additional knowledge, our skepticism is directed at the reportage and not at the subjects themselves.

WRONG NUMBERS

Miscounting

How easy it is to collect or record numbers incorrectly! In any complex data collection, many people are involved. Not all "get the word" and not all are capable of carrying out the instructions. In addition, it is not unusual for data collection and recording to take place in a work environment where the people are required to carry out several tasks at once, and data may get mangled and abused.

Policing the Police

Arrest statistics are crime's basic data, and producing these statistics dramatically illustrates the difficulties of data collection and recording. Considerable ingenuity and effort have been applied to the problem of arrest statistics.

Unfortunately, ambiguity in the definition of an arrest has been a continuing problem. After a long history of judicial and statistical concern for just what is an "arrest," arrest-counting rules have been established and widely disseminated through training as a part of the Uniform Crime Reporting (UCR) program. The rules are simple enough. For example, "Count one (arrest) for each person (no matter how many offenses he is charged with) on each separate occasion (day) he is arrested or charged" (11, as quoted in 12).

Skeptics have suggested that police departments increase or decrease the reported number of arrests in response to political pressure. The argument is made, and supported in some cases, that the greater the number of arrests, the better the police agency is doing its job. The major decision affected by arrest statistics is which policy to use to deter crime: aggressive police patrol practices, or more arrests.

But a study by Sherman and Glick for the Police Foundation shows that police departments underreport as well as overreport the number of arrests (12, p. 27). Sherman and Glick give an excellent summary of the problems of getting correct numbers for arrests; their observations have application to many fields.

Glick spent two weeks in each of four police departments, auditing the departmental counts for a month. He found counts of total arrests by department in error by -2%, $+5\%$, -5%, and $+14\%$. The error in count for all four departments combined was $+1.5\%$ out of 3584 actual arrests.

A sample of four does not allow us to make meaningful projections to all police departments. But it does suggest that serious miscounts can exist. Can we take it for granted that individual errors in all the reporting police departments in the country will balance out in every reporting period? What are the consequences of local policing decisions based on count errors as great as 14%?

Miscounting Is a Statistical Crime

In October 1997, the FBI announced that they were deleting all of the crime statistics reported by the city of Philadelphia for the preceding 18 months. The flaws in the data were discovered at a hearing on policing at which the Mayor of Philadelphia claimed a 17% reduction in crime in the first half of 1997 as compared to the preceding last half of 1996:

> Analysis of [the Mayor's] numbers showed that police counting methods did not conform to guidelines. Instead of reporting crimes that actually occurred in a given period, the police were counting the crimes as they logged them into their records [13].

The Philadelphia police are recounting their crimes, but in the meantime, Philadelphia will be absent from the FBI's published data on crime for the first half of 1997.

Where Are the Bean Counters When We Need Them?

In March 1997, the U.S. Department of Agriculture announced that an error in counting the stock of soybeans in Mississippi led to a nine-year high in the bushel price of soybeans in the U.S. (14). Billions of bushels of beans are involved, and consequently, the additional cost of soybeans will be hundreds of millions of dollars.

But even more money was at stake in the early 1990s bailout of the savings-and-loan institutions after the disastrous crash of the 1980s. About $100 billion cannot be found, according to the General Accounting Office:

> [The GAO] reports that the Resolution Trust Corporation (R.T.C.), which oversees the bailout, has not only lost track of billions of dollars in loans, real estate, and other assets seized from failed institutions, but also has lost track of the cash it got when those assets were sold. . . . In one instance, the GAO said, $7 billion appears to have vanished from the Government's books, but it does not know whether that is due to fraud, incompetence or simply poor record-keeping. The R.T.C. acknowledges that it has problems [15].

The article was reporting a request to Congress by Secretary Brady for a $55 billion bailout. Perhaps he should have looked under a couple of mattresses.

It's one thing to miscount, another to not even try to count. In yet another argument over education:

> CUNY [City University of New York] officials contend that many community college students go on to senior colleges and earn degrees, but they cannot supply the figures. CUNY does not track the number of students who transfer from community colleges to four-year colleges, and graduate [16].

And this was in connection with a debate over the need for, and value of, remedial math courses! Teacher, teach thyself.

Misclassifying

When a scientist classifies an object for statistical purposes, other qualified scientists should classify it in the same way. We call this type of verification "the replicability of observational units." In some scientific fields (e.g., geology, physical anthropology, archeology), the failure of replicability is called "observational discrepancy." Paul Fish studied observational discrepancy in archeological classification and found "provocative and some surprising results" [sic] (17).

In one location, a single instructor trained experienced analysts in the taxonomic scheme for classifying one of the most clearly defined types of ceramic shards (pieces of pottery). Despite this, the proportion of discrepancies in classification of individual shards within the taxonomic scheme ranged from 22% to over 30% between any two analysts. The discrepancies were distributed almost evenly among the analysts and the types of items being classified. In some cases, analysts differed in the definitions of characteristics. In others, the same analyst changed definitions from one time to another. Even when classification was based on characteristics measured with simple instruments, the analysts occasionally disagreed.

Biased Against Data

Connecticut has a state law that requires local police to record and classify all crimes motivated by bigotry and bias. The police are then to give the data to the State Police. The purpose of the bill is to enable Connecticut's legislature to provide information about the prevalence of bias crimes to the sponsors who want this information in considering new legislation. Since the goal is to know the number of incidents needed to take legal action, you certainly want to count every incident.

Unfortunately, that is exactly what is not happening (18). The *reported* number of incidents is known to be too low. Unlike many other incidents in which undercounting takes place, there is not just one, but several "smoking guns." The public information officer for the New Haven Police Department "conceded that his department did not report bias crimes" to the state police because of the amount of time it took. An advocate for homosexuals said that most police departments do not report these crimes, and that "They blatantly tell us they don't have the time. And most people don't even know [about the requirement to report]." The director of police services at the University of Connecticut said that he felt that university police departments tended to report bias crimes at a higher rate than other police departments because "campuses are more sensitive to this sort of thing. . . . Our statistics are going to be high, higher than those who don't take it as seriously as we do."

But unfortunately, his department's data are aggregated with those from the other departments, and a serious, unknown undercount exists. What kind of legislative process can we expect based on these poor-quality data?

BAD MEASUREMENT

One Way or Round Trip?

Years ago, one of us received a questionnaire asking "How many miles do you commute each day?" The question was confusing. Did it mean one way or round trip? There was considerable doubt in the minds of typical respondents, but it is not reasonable to assume that potential respondents are going to make a phone call (perhaps long distance) to ask the survey director what was meant. We assume that most respondents made the assumption that one or the other was intended and gave a number.

If the originator wanted the one-way commuting mileage, then the value estimated from this survey would be high, because some respondents answered with the round-trip mileage. If the originator wanted the round-trip mileage, then the estimated value would be low, because some of the respondents answered with the one-way mileage. Either way, the numbers generated were useless.

Will the Real Number Please Stand Up?

The United States' Original Inhabitants

How correct are the numbers appearing in the reports of the U.S. Census? Some may be as correct as the human mind can devise. Others must be taken with a degree of skepticism. We illustrate the application of skepticism to a particular case, the number of Native Americans. The U.S. Census reports the number of Native Americans to the last digit as though it is correct to the last person, but we round off to tens of thousands to avoid giving a spurious impression of precision. Table 5.1 shows the numbers and annual percentage changes for 1600 to 1990.

How can we check these numbers? One way is to look at the percentage change in population from decade to decade. This is easier with the American Indian population than with others, since this group increases in number almost solely by an excess of births over deaths; there is virtually no immigration of Native Americans from foreign countries. A small number of Cana-

Table 5.1 Estimated Number of Native Americans in the U.S.: 1600 to 1990 (in Thousands)

Year	Numbers in thousands	Percent change per year
1600	1000	
1820	620	−0.2%
1850	450	−1.1%
1870	330	−1.5%
1890	270	−1.0%
1920	270	0.0%
1950	360	1.0%
1960	520	3.7%
1970	750	3.7%
1980	1480	7.0%
1990	1940	2.7%

Sources: Ref. 19, p. 108; Ref. 20, Table 52.

dian Native Americans may cross into the United States, and a small number of U.S. Native Americans may cross into Canada. No one knows how many make this transition, but in the absence of a movement large enough to be reported as a special event (which has not happened in the past three decades), the number of immigrants can reasonably be assumed to be small.

The maximum possible annual increase in a population due to the excess of births over deaths during a decade is 2.7% to 3.5%. The annual increase from 1950 to 1970 was 3.7%, which is high enough to arouse suspicion. As Table 5.1 shows, the annual increase from 1970 to 1980 was a phenomenal 7%, and from 1980 and 1990 a most reasonable 2.7%. But the rates of increase attributable to the excess of births over deaths for the period 1960 to 1980 are impossibly high.

At least one number is wrong—but which? One of us (Jaffe) investigated this situation in detail (19, pp. 115–116). We believe that the explanation lies in the quality of the basic data.

Prior to World War II, the U.S. Bureau of Indian Affairs and the U.S. Bureau of the Census separately reported counts that were remarkably close (for example, see 21, Table 8). However, after World War II, the Census reported significantly larger numbers of Native Americans than the Bureau of Indian Affairs. A.M. Gibson, of the University of Oklahoma, in his book on the American Indian (22), describes in detail the social changes which could account for U.S. Census Bureau counts of the Native Americans rising dramatically.

Starting in 1958, federal U.S. policies toward the Native Americans were softened, and under President Kennedy rights were restored, and support for improved social services led to genuine changes. In addition, the climate of tolerance toward minorities greatly improved. One index of this is the increased number of marriages between Native Americans and whites (19, p. 116). Since the U.S. Census relies on ethnic self-identification, the increasing social acceptability of ethnicity in recent decades, and an increase in concern for one's "roots," more people who have some Indian ancestry (in the case of the Kaw tribe, as little as one part in 64, according to Tony Hillerman's *New York Times* Op-Ed article [23]) may be changing their identity to American Indian from one census to another. No one knows for certain, and only a suitable survey holds even a promise—let alone a guarantee—of an answer.

The U.S. Mainland's Newest Inhabitants

The migration of Puerto Ricans to New York after World War II set off another round in the "number, number, who's got the number?" game. As Clarence Senior described the situation:

> With the end of World War II, plane loads of Puerto Ricans began arriving in New York City in response to the employment opportunities which

the city offered . . . and they are facing the same social and economic problems as the earlier immigrants [24, pp. 1–2].

New immigrants always seem to pose problems for the receiving communities, and the situation with regard to the Puerto Ricans was no different. The remarkable similarity of the public reaction to each new immigrant group helps us to understand these continuing problems. Both official New York sources and the media greatly exaggerated the numbers of Puerto Rican immigrants during the late 1940s:

> Prior to 1948 . . . various estimates placed the Puerto Rican population of New York City at half a million and upwards. In 1948 a Columbia University study [by Senior] showed there were not more than 200,000 Puerto Ricans in New York City . . . if that many [24].

The U.S. Census for 1950 reported 245,000 Puerto Ricans in New York City. In the two years between the Columbia University estimate and the Census of 1950, some 50,000 migrants from Puerto Rico to the mainland were reported; the Columbia University study's estimate was verified by these official numbers. Would the average newspaper reader have been aware of the exaggeration in newspaper stories which reported double the actual number of migrants?

Even in the Best of Circumstances . . .

The preceding examples pertain to the measuring instrument—to making sure that the data collection does not introduce flaws in the data available to the investigator. But we also have problems at the respondent's end. Even with the best of intentions and the clearest definitions of the required measurement, you can get bad data.

In statistics courses that one of us (H. Spirer) taught to Master of Business Administration (MBA) students over 25 years, he used a class exercise designed to illustrate data collection and analysis. The basic data were the students' incomes. The exercise was as follows: The instructor suggested that the class might be interested in the income of the group as a whole. The class was allowed to discuss "income" in depth, raising such questions as: "Why do we want to know?"; "What kind of income do we want to deal with (i.e., salary, salary plus bonus, salary plus outside income from investments, contributions from a spouse)?"; and so forth. The discussion continued until the class reached consensus. In the many years this exercise was used, the decision was almost always to collect data on salary only and to prorate part-time income to a full-time equivalent.

With the general goal settled, the discussion turned to how to collect the data. Usually, the class decided to collect the "annual salary" (actual or prorated) as the basic data. The class then had to decide on the exact format of the data to be collected. Should it be expressed as annual salary in dollars? With or without comma for thousands? Annual salary in thousands? With or without a "K" to indicate thousands? Leading zeros? Dollar signs? How are the unemployed to indicate their situation? Is there a special indication for part-time workers?

The exact format of the response was determined by consensus. Then index cards were distributed and the class members put their responses on them in accordance with the rules just determined. The instructor collected the cards and the responses were tabulated. The average class size was 30 and there were never fewer than three defective responses in any class!

Some respondents gave their annual salary in dollars (with and without commas), or gave their weekly salary in dollars or thousands, or spelled out the value, or gave two values (always in violation of the rules just agreed to), and so forth.

What is really remarkable here is that this is the result when the respondent was also the designer and data collector, the definition of what was to be collected had been discussed for more than 30 minutes, there was no pressure on the respondent, the respondent had been able to raise any issues of ambiguity and have them resolved, and the respondents were graduate students with business experience. Imagine what kinds of data are collected when the question is asked with little if any explanation and possibly in syntactically incorrect form, or when the respondent isn't interested in the survey. That is the situation in all too many cases.

Even when the question is clear-cut, explicit, and well stated, and all reasonable precautions are made to avoid erroneous response, there are questions for which answers cannot be accurate. Much so-called market research uses questions such as: "How many pairs of socks did you buy this year?"; "How many times has your dishwasher been repaired since you purchased it?"; "What is the average amount you spend on a restaurant meal?"; and "How much money do you have in your checking account?" Setting aside all questions of confusion about the nature of the answer, bias, or desire for secrecy, most people do not have the mental storage and recall capacity to answer these questions without looking up the information, and while the size of one's checking account may be accessible, how many people keep records of purchases such as socks? How close to garbage are the answers given to such questions, which then are analyzed at great length using complex statistical procedures and ultimately offered to the public as "real" information?

Underdeveloped Birth Rates in Developing Countries

As we have discussed, we immediately suspect that the underlying data are of low quality if there is disagreement among several statistics purporting to measure the identical concept. These discrepancies are often clues that the

Table 5.2 Birth Rate per 1000 of Population

Country	Year	Estimate		Source	
		High	Low	High	Low
Brazil	1965	45	38	d	e
	1974	39	36	c	d
Nicaragua	1965	49	44	e	a
	1974	48.3	44.5	a	b
Egypt	1965	42.5	41.1	f	d
	1974	38	35	e	c

Sources: (a) United Nations Statistical Office; (b) U.S. Bureau of the Census;
(c) The Population Council; (d) United Nations Population Division;
(e) USAID Office of Population; (f) the Population Reference Bureau.

reported numbers are guesses and not scientifically derived estimates. Here we illustrate this point with statistics purporting to show changes in the birth rates of developing nations, an issue of political and social importance. The authors of an article published in *The New York Times* have no doubts. They find that "Survey Reports Fertility Levels Plummet in Developing Nations" and make no mention of conflicts in statistics on birth rate (25).

Table 5.2 shows some of the published birth rates for Brazil, Nicaragua, and Egypt, as assembled and analyzed by Joseph Cavanaugh (26). We can calculate high and low limits for the estimates of changes in birth rate between 1965 and 1974 using combinations of the extreme values of the birth rate estimates. For example, for Brazil, we could estimate the birth rate in 1974 as being 80% of that in 1965 by dividing the low 1974 birth rate estimate—36—by the high 1965 birth rate estimate—45. Or, for the same time and country, we could estimate the birth rate in 1974 as being 103% of that in 1965 by dividing the high 1974 birth rate estimate—38—by the low 1965 birth rate—39—estimate. Thus the birth rate change estimates for Brazil could range from a decrease of 20% to an increase of 3%. Table 5.3 shows the results for these three countries.

Table 5.3 Range of Estimates of Percent Change
in Birth Rate, 1965 to 1974

Country	Range
Brazil	20% *decrease* to 3% *increase*
Nicaragua	9% *decrease* to 10% *increase*
Egypt	18% *decrease* to 8% *decrease*

What is the "correct" change in the birth rate? Joseph Cavanaugh showed similar figures for 30 countries, and the variability and conflict illustrated above were present in all these countries.

It is a misuse of statistics to use whichever set of statistics suits the purpose at hand and ignore the conflicting sets and the implications of the conflicts. We know very little about birth rates in many countries, and this should surprise no one. If you look at the difficulties we have experienced enumerating the large and diverse population in our own country despite our substantial resources, what can we expect in less wealthy societies? How many unrecorded births take place daily in countries?

Did the originator of the report or the writer of the newspaper story have a secret wish (conscious or unconscious) that the birth rates should go down? Or just bad data?

THE UNKNOWABLE

You can never obtain some kinds of data. For example, there is a great deal of interest in the "underground economy" of the United States because the monetary transactions in this economic sector are not recorded and cannot, therefore, be used for tax purposes. This lack of information concerning transactions also concerns economists because it affects their ability to estimate economic variables that are important to economic decision making. Criminal justice issues are involved as well. There is considerable interest in tracking "laundered" money from drug traffic profits, for example, and some recently announced bank regulations were formulated to attempt to do such tracking.

Some economists argue that one way of estimating the amount of money moving in the underground economy is to observe the flow of money into financial institutions (27). But this only represents, at best, the money *saved* in financial institutions. Since underground *spending* cannot be estimated, it is difficult to estimate the savings rate for the nation as a whole with this one important sector unknown. Any unsubstantiated estimates should be viewed with suspicion.

DISAPPEARING DATA

In most cases, the best way to get statistical evidence for cause-and-effect relationships in human behavior is to follow specific individuals over a long period of time. Researchers usually do this with longitudinal surveys, reinterviewing (or reexamining) the individuals at regular intervals (day, month, year) for the period of the study. Without a longitudinal survey, data on the past (feelings, actions, and so forth) can only be obtained from present memories. Such memories are clouded by the deficiencies in the individual human memory and selectively recalled and modified according to current interpretations.

Getting the True Dope

Drug use (and abuse) is a major social problem in the world. There is a great need for public and private policy and direction concerning cause, cure, and control. To be effective, such policy should be based on a clear understanding of the underlying mechanisms. We seek such understanding from surveys as well as from experiments. But because of the ethical issues and the long-term nature of drug use and abuse phenomena, what we can learn from experiments is limited.

Many researchers have started (and some have completed) longitudinal studies of drug use and abuse (28–30). Valuable information has been gained, but:

> A major problem [is] . . . the differences between those persons who complete questionnaires or are interviewed in all waves of a panel study and those who are captured in the first wave but are lost in later ones. . . . Of those seen in the first wave, only a percentage are seen in the later ones. It is usually only those who furnish data in all waves on whom data analysis is based, but if a large percentage of the original random sample is lost the generality of the conclusions becomes questionable [29].

Just how many of the original starters were lost? For three studies (28, as quoted in 29) the proportions of original starters present in the final wave were 73%, 66%, and 44%.

Do we have reason to believe that dropouts from these longitudinal studies were random occurrences and therefore do not affect the conclusions? Alas, no. For "It was precisely the drug users, poorer students, and truants who were lost. . . . Data on . . . [other] percentages lost . . . also indicate that it is lower-class, minority-group members who are most likely to be lost" (29).

This does not mean that the data on the survivors in the final wave are invalid. What it does mean is that the data for them cannot be generalized to the original target population as a whole, that some statistical methods have little validity, and that, since data are missing for the most affected groups, the research may not fulfill the original purpose of the survey.

Jobs for Youth

Another study was concerned with the possible effect of initial job experiences on subsequent employment: "The premise . . . is that in the early career of a young person's life, initial job experiences and attitudes are critical in shaping ultimate unemployment experience" (31, p. 2). To test this premise, the investigator analyzed the age 14 to 24 cohorts of the National Longitudinal Survey, following males from 1966 to 1975 and females from 1968 to 1975.

Losses from the initial sample were considerable. As well as we can estimate from the information given (31, p. 37), about 25% of the male respondents were lost in the first three years. It is impossible to determine how many disappeared from the study in the following six years. No loss information is available for females. This does not mean that the researchers do not know how many subjects were lost, only that in their published paper they do not explicitly tell us how many were lost and do not give us a basis for judging the accuracy of their conclusions.

In this case, we don't know (a) the proportion of male and female survivors to the end of the study, or (b) their characteristics (as we know in the case of the drug survey just discussed). Thus, we have no basis for judging the applicability of the results. The author carried out considerable data manipulation, but no amount of computation can make up for the deficiencies in the basic data.

MORE IS LESS

A Catch-22 for Economic Forecasters

Some processes defy logic. Economic forecasters have a tough job at best. They have to use their knowledge of the current state of the economy with reference to its past history to predict the future. For their purposes, the "state of the economy" is primarily determined by about 30 statistical measures of economic activity—the trade deficit, manufacturers' shipments, wages, employment, and so forth. If their data are of high quality, then they will know with some reasonable precision the values and trends of their 30 or so statistical indicators. From this, they still have the difficult job of trying to divine the future. To do this, most economists employ fairly sophisticated analytical models for which they use computers, which makes the process somewhat easier. But we should not be surprised that they differ, although they tend to make generally similar forecasts. Why does this happen, since they use basically the same data?

The process itself is shaky. "We are all so eager to signal changes in the economy's direction that we jump to conclusions based on very tentative information," says one forecaster (32). Unfortunately, the data are worse than "tentative." Another forecaster lays it on the line; "We are making forecasts with bad numbers, but the bad numbers are all that we've got" (32). In most other professions this would be an absurdity. Imagine a dentist telling you that he is filling the cavity in your tooth with wax, because that is all he has.

A large part of the problem lies in the collection of the data. Since there is such demand for immediate results, incomplete data are reported and, in subsequent weeks or months, are updated as corrections and additional data are received. You can easily see this phenomenon at work by looking up the value of the *same* statistic for the *same* year in successive editions of the *Statistical Abstracts of the U.S.* You will see in many cases—especially economic statistics—that the value given for the same statistic for the same year is different, depending on which issue of the *Abstract* you use. For example, in the 1996 *Statistical Abstract,* the total value of the 1987 national farm inventory of sheep and lambs is $799 million (19, Table 1114). In the earlier, 1990 *Statistical Abstract,* the value of the same inventory is given as $782 million (33, Table 1162).

But the economists' problem is that "When you are advising investors, you have to do it with conviction" (32). Good sense would indicate that forecasters should revise their forecasts as they get more, better data. But good sense does not always operate well in the presence of bad data. "You lose credibility if you revise your forecast too frequently," says another forecaster. And there's the catch. A forecaster makes a forecast based on bad data but has to be careful about revising that forecast when he has better data, because then people wouldn't believe his next set of estimates. But the first estimate is based on bad data. Our forecaster is surely impaled on the horns of a serious dilemma.

Forecasting the economy is a complex process, and the forecasts themselves enter into shaping the future. But should we really use bad data because they are all we have and then not improve our forecasts when we have better data just because the users would think that you had made a poor estimate in the first place? Our heads are spinning.

SUMMARY

Our Recommendations

The single greatest lesson you can take from these examples is this: no matter what your role, be it writer, reader, researcher, student, or teacher, or any combination, you must always be skeptical.

For the Reader

1. If data or results based on data are given, look for a source in the article or report. If you can't find one, and you are more than a casual reader, try to get

information from the author. If the author's source is an article or report, look it up.

If you find no source and cannot get one from the author, look for data which will confirm or discredit the data. If your best efforts lead you to believe the data are mythical, unknowable, or unknown, then draw your own conclusions accordingly.

2. Understand that even the best of researchers cannot always collect perfect data. A simple counting exercise can be flawed; accept this fact but remain skeptical.

3. Never accept extrapolating results to a whole population if the research methodology obviously was based on inadequate or inappropriate data collection.

For Authors

Give sources when you give data or results of analyses. Don't base headlines and conclusions on mythical, unknowable, or unknown data. Help the reader by giving enough information about the data collection process to enable the reader to judge the results. Here are some suggestions to help you avoid the serious misuses which result from poor data:

1. When planning to do a survey, be sure you know the true "purpose": the question to ask is: "What do I want to know?" When you can answer this question, you will ask questions that elicit the appropriate information.

2. Understand that you cannot always collect perfect data; make allowances for this (a simple counting exercise can be flawed!), and be honest with your readers about it.

3. Always check your sources.

4. Check for changes in definition which might change your analysis of the data.

5. If you see any suggestion of conflicting data, do not ignore it— even if you like the results *your* data give.

6. Never generalize your results to a whole population if you have inadequate or inappropriate data collection.

7. Learn how to properly aggregate and disaggregate.

Are We Asking Too Much?

We don't believe that we are asking too much of either readers or authors. Not everyone agrees with us, however. The following is a criticism of the preceding recommendations for authors from a statistician:

> It should be recognized that newspaper articles exist under restrictions
> of space. Also, while I don't condone the way the media often distort

reports to make them interesting, I think it's obvious that if they sur-
rounded everything with scholarly heaps of source references and discussions of procedures they would lose most of their readers [34].

We respectfully wish to differ. As we note several times in this book,
when *The New York Times* reports on a sample survey, it gives a description
of the sampling methodology, the magnitude of the sampling error, and a brief
discussion of the meaning of the relevant technical terms. When *Redbook*
reported the results of a survey of sexual attitudes, the editors made space for
a caution about the self-selected nature of the sample (35).

It is easy to use statistics correctly when you want to! In a letter to the
editor in *The New York Times* (a type of newspaper item in which space is at a
great premium), the secretary-treasurer of the American Philological Association, Roger Bagnall, finds it possible to be thorough about naming sources.
We excerpt the relevant paragraph of his letter and italicize selectively to
show how you can meet our standards for avoiding the misuse of statistics
that comes from not giving source information:

> The output of new Ph.D.s in classics has fallen by about half in the past
> decade, *according to the National Research Council: from 88 in 1974*
> *and 93 in 1975 to just 44 last year.* By contrast, the number of jobs ad-
> vertised through the *American Philological Association's placement ser-*
> *vice has increased in the past few years to last year's (1984–85) 98 firm*
> *and 29 possible positions.* Not all of these are tenure-track, but 40 of the
> positions advertised last year were tenure-track *according to the copy of*
> *the advertisement* [36].

It can be done if the will is there. Do it!

6
Graphics and Presentation

Look here, upon this picture, and on this.
                                                    —Shakespeare

*And shall shew signs and wonders, that they may lead astray, if possible, the
elect.*
                                                    —Matthew 24:24

INTRODUCTION

To the ancient Chinese proverb "one picture is worth 10,000 words," we add
"for good or for bad." A colorful, dramatic graph can have a far greater impact than an uninspiring table of numerical values, for as the famous statistician John Tukey said, graphs enable us to "notice what we should have seen
in the first place." Graphs are an excellent statistical tool for presenting results in certain circumstances, especially to nonstatistical audiences, but the
choice of presentation method is demanding, and we see too many misleading or incorrect graphic illustrations.

   The use of personal computers makes it quite easy to choose from a
wide selection of graph styles. The formating capabilities of modern statistical programs (such as SPSS, SAS, MINITAB, DATADESK, and many others) as well as "spreadsheet" programs (EXCEL, LOTUS, QUATTRO) offer
a myriad of choices to the user. For example, Microsoft EXCEL 7.0 gives the
user a choice of 15 types of charts. Each chart selection has between four and
10 different standardized formats. The total number of combinations of styles
and standardized formats is 1530! It is also possible for the user to infinitely
modify most of the characteristics of the charts.

   Thus, the user can, with a few key strokes, instantly change the nature
of a graph and display the data in a new way. More importantly, the user also
has the ability to mislead both self and others, by dynamically altering the

appearance of the plotted data. Was this possible with pen and pencil? Of course! But today, it is so easy to get so many different effects.

Edward Tufte, William S. Cleveland, and Howard Wainer have given professional statisticians and nonstatisticians many new insights into this process [1–3], with both good and bad examples. In this chapter, we look at several actual cases in the hope that we will all learn to reduce the number of graphical misuses which can have such pernicious effects on our thinking.

SOME HORROR PICTURES

Dr. Nowall, our fictitious researcher, cleverly includes graphs in his reports and studies wherever and whenever he can, which is generally good practice.

He has discovered that sober, restrained reports on his findings, regardless of their importance, rarely bring him requests for public statements or appearances. However, if he can announce a threat of disaster, or some apparently startling change based on statistics, these will get him front-page publicity and invitations to appear on TV talk shows. He has learned that no words are as dearly beloved of the headline writer as "soaring" and its counterpart, "diving." And the graph is the way to go.

If he can plot his data to give the impression of a "soar" or a "dive," attention will be paid. Even if the threatening quantity is really merely creeping upwards or downwards, or just varying within its normal range, he may be able find a way to plot it so that it soars or dives on the graph, striking fear into the hearts of the readers. Let us look at some examples of the technique.

Sinning Against the Vertical Axis: The Missing Zero and the Missing Baseline

The Great Land Boom in the Suburbs of Los Angeles

In terms of current dollars, real estate prices in the U.S. rose steadily in the 1970s. The general feeling was one of an exhilarating upward-bound ride, and readers undoubtedly were impressed when the lead paragraph of a 1978 *Barron's* article called it "the biggest land boom in American history" (4). To show the dramatic growth in land values, the article included the time series of Orange County home prices, shown in Figure 6.1. The growth in values from April 1970 to October 1977 appears to be truly outstanding. At first glance, it looks as though there has been an increase of about 40-fold during this period since the ratio of the height of the line plotted for October 1977 is about 40 times that of the height plotted for April 1970. This effect occurs because all the values below the real estate index value of 140 are deleted,

ORANGE COUNTY HOME PRICES
April 1970–October 1977

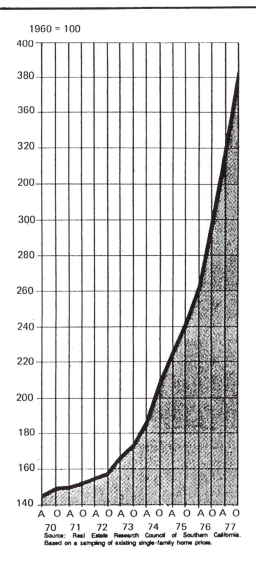

1960 = 100

Figure 6.1 "Soaring" home prices in Orange County. Soaring is emphasized by the *sin of the vertical axis*, no return to zero or indication of the absence of zero on the vertical axis. The designations "A" and "O" along the horizontal axis denote "April" and "October," respectively. (From Ref. 4.)

greatly exaggerating the apparent ratio in values. When the graph is redrawn (Figure 6.2) with the left-hand scale carried down to zero, you can see that the true ratio of growth is about 2.6, not 40. Not a bad gain in land values over a 7.5-year time period—all of us should be so lucky—but a rise of 2.6 times is nowhere near as dramatic as the published graph showing an apparent 40-times growth based on the missing zero on the vertical axis. We call this the *sin of the missing zero.*

It is a major statistical sin to show a graph displaying a variable as a function of time with the vertical (left-hand) scale cut short so that it does not go down to zero, without drawing attention to this fact. This sin can create a seriously misleading impression, and, as they do with most sins, sinners commit it again and again. Read today's newspapers, magazines, and reports, and you will see many economic and demographic time series that are drawn in this misleading way, with no caution to guide the viewer.

Twisting on the Axis

Figures 6.1 (sin of the missing zero) and 6.2 (redrawn with zero on vertical axis) are also examples of another sin. In Figure 6.2, the addition of the zero on the vertical axis does give a more realistic view of the growth in the Orange County home prices, but the chart still has a dramatic soar. In large part, the soaring visual effect occurs because the vertical axis is an "arithmetic scale." Using this scale, an increase of 10% is twice as long on the vertical axis for a base value of 300 as it is for a base value of 150. Thus, numbers that are increasing by constant percentages appear to soar.

Important numbers such as salaries, prices, costs, the inflation rate, populations, and victims of epidemic diseases tend to grow by annual percentage amounts. Figure 6.3 shows what a constant percentage change of 14% a year looks like over a 10-year period with an arithmetic vertical scale; this is a soar in the making.

If you want to show the growth of numbers which tend to grow by percentages, plot them on a *logarithmic* vertical scale. When plotted against a logarithmic vertical axis, equal percentage changes take up equal distances on the vertical axis. Thus, a constant annual percentage rate of change will plot as a straight line. The vertical scale on a logarithmic chart does not start at zero, as it shows the *ratio* of values (in this case, land values), and dividing by zero is impossible.

You can see this effect in Figure 6.4, which illustrates what the constant percentage growth of 14% looks like when plotted against a logarithmic vertical axis. We plot the average rate of inflation (the Consumer Price Index for All Urban Consumers) and the Orange Country home price index against a logarithmic vertical scale in Figure 6.5. Here, you can see clearly the behavior

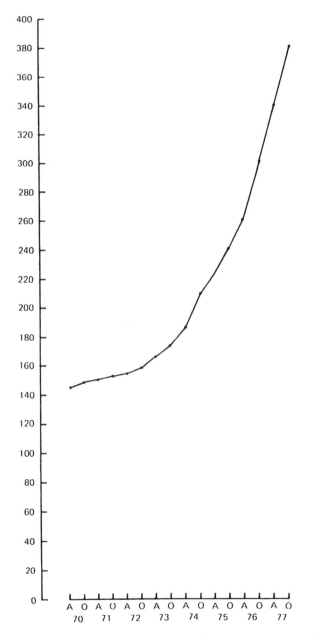

Figure 6.2 Figure 6.1 redrawn with the zero shown on the vertical axis. The magnitude of soaring is clear because you can now see the value of the index in April 1970.

14% Growth on Arithmetic Scale

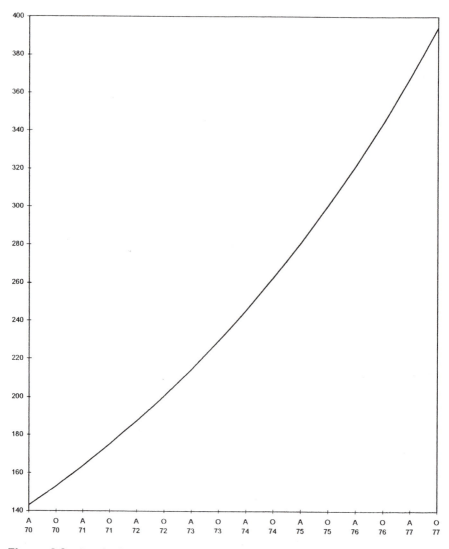

Figure 6.3 Graph of a 14% per year growth during the period April 1970 (A 70) to October 1977 (O 77) plotted to an arithmetic scale on the vertical axis. The vertical axis might be any variable of interest, such as home prices, a real estate index, pay, number of housing starts. See how it "soars."

14% Growth on a Logarithmic Scale

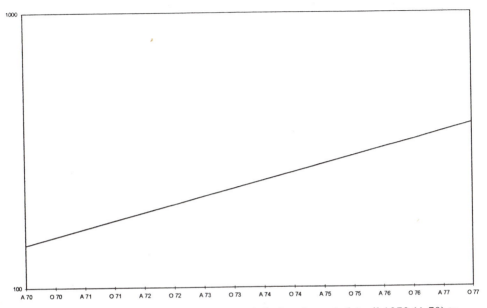

Figure 6.4 Graph of a 14% per year growth during the period April 1970 (A 70) to October 1977 (O 77) plotted to a logarithmic scale on the vertical axis. Note that the growth that appeared to soar in Figure 6.3 is now an upward-trending straight line.

of the home prices; they grow at a rate less than inflation from 1970 to 1974, and then at a higher, relatively constant rate of about 14% per year for the next three years.

Comparing the rise in home prices with a measure of the overall rate of inflation such as the Consumer Price Index is one way to look at these data. Another way is to compare the change in home prices to the change in family income, because homes are paid for out of family income. Unfortunately, we lack information on family income for Orange County for the years shown in Figure 6.1 and cannot make this comparison. However, when you see such a chart again, you should ask: What were the changes in earnings? In family income? In national income? It is important to compare the raw data with some measure of economic change that affects the home buyer.

Sinning Against the Horizontal Axis

The Japanese Are Coming

So many important economic variables seemed to be soaring in the 1970s and '80s! In the preceding case of the great land boom, we saw a soar to delight

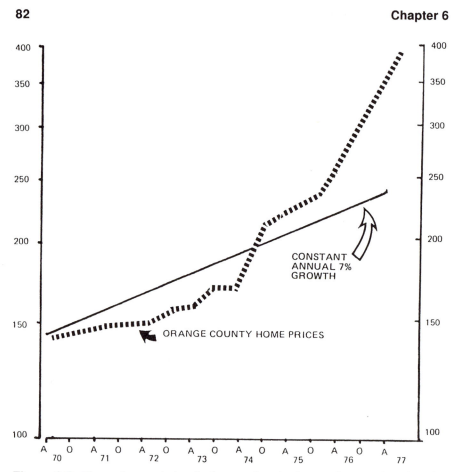

Figure 6.5 The real estate index for home prices in Orange County plotted on the ratio scale graph of Figure 6.4. The soar is now revealed as a nearly constant growth at an annual rate lower than the average inflation rate until 1974, when it climbs at a more rapid rate. This graph clearly shows that during the period of interest, the average annual percentage rate of increase of the index is twice that of the Consumer Price Index.

the lucky early purchasers of Orange County real estate. Other soars were perceived to be less positive in their effect. In Figure 6.6, you can see the projected growth of Japan's multinational corporations distorted into a threatening "soar" by a manipulation of the horizontal scale (6). The sins of the horizontal axis are less frequent, but just as sinful as those of the vertical axis. Not surprisingly, the misleading graph of Figure 6.6 was published at a time when there was great public concern that Japanese multinational corporations were gaining undue prominence.

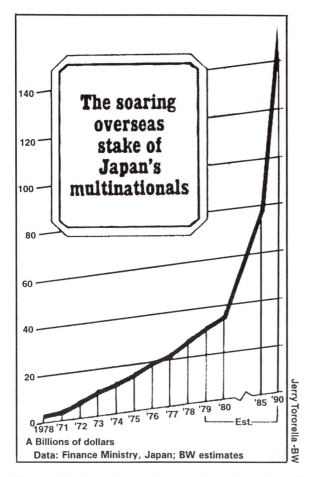

The soaring overseas stake of Japan's multinationals

A Billions of dollars

Data: Finance Ministry, Japan; BW estimates

Jerry Tororella -BW

Figure 6.6 Two changes in the scale on the horizontal axis (1980–1985 and 1985–1990) give a "soar" to Japan's overseas stakes. Although one of the changes in scale is announced by a wiggle in the horizontal axis (between 1980 and 1985), the result is a serious misuse. (From Ref. 6.)

On the original chart, the decade from 1970 to 1980 gets a horizontal distance of three times that given to the decade from 1980 to 1990. The effect magnifies the apparent "soaring." To make sure that you got a strong soaring feeling, the graph's designer also exaggerated the effect by tipping the horizontal scale upwards on the right side.

Look carefully and you will see how this graph misleads the viewer. During the first 10 years (1970 to 1980) shown on the graph, Japan's "overseas stake" grew about 10 times. During the following 10 years (1980 to 1990), Japan's "overseas stake" is projected to grow only 4.5 times. But the way the

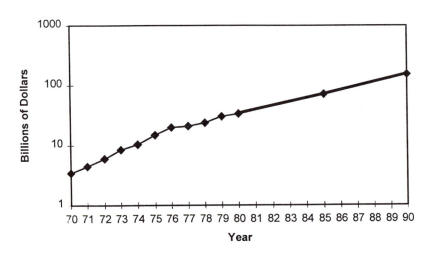

Figure 6.7   Figure 6.6 redrawn with a consistent scale on the horizontal axis and a logarithmic scale on the vertical axis. The annual growth rate is highest from 1970 to 1976 (about 35%), is erratic from 1976 to 1980, but the rate for the decade 1970 to 1980 is 25%. From 1980 to 1990 the annual growth rate is nearly constant at 15%.

graph is drawn, this smaller growth in the second decade appears to soar upward. The average annual growth rate during the first decade is about 25% per year, and during the second, "soaring" decade, the average growth rate is about 16% per year.

Thus, the threefold compression of the last 10 years (1980–1990) on the horizontal scale has made a lower growth rate (16%) loom larger than a higher rate (25%). A clear picture of the projection of Japan's overseas stakes is shown in Figure 6.7, which is a plot of the graph of Figure 6.6 on a consistent horizontal scale and with a logarithmic vertical axis. You can now see that Japan's overseas stake was indeed increasing, but at a declining rate. We can only wonder whether the originator of this graph knew what he or she was doing. Can you find it in your heart to forgive this deliberate or inadvertent sinner?

When Graphical Methods Conceal Rather Than Reveal

For all its sins, the plot of the Orange County home price index of Figure 6.1 does not conceal the relationship which it distorts. It is a straightforward line chart of a time series, where the value of a variable is plotted as a point on the

Table 6.1 Estimated Power Consumption (Quadrillions of BTUs) for 1975–1995 and Predictions for 2005, by Source.

Year	Hydro	Nuclear	Gas	Petroleum	Coal
1975	3	2	20	33	13
1985	3	4	18	31	17
1995	3	7	22	35	20
2005	7	48	32	70	30

vertical axis at a place on the horizontal axis corresponding to the year. To give a feeling for the progression of time, the successive points are connected with straight lines.

We are indebted to Howard Wainer of the Educational Testing Service for bringing to our attention the type of graphical confusion that results when the straightforward approach is not used (7). Table 6.1 shows the estimated power consumption by source in quadrillions of BTUs for the years 1975, 1980, 1985, and 1995, and the predictions for 2005 (8).

In 1990, these data were presented in the chart similar to that shown in Figure 6.8. Suppose that you were asked, as were 17-year-olds taking one form of the test for the *National Assessment of Educational Progress*, to de-

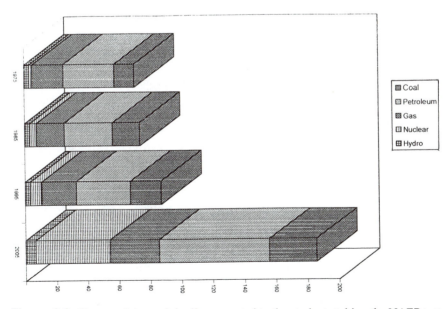

Figure 6.8 Energy plot as originally presented to the students taking the NAEP test.

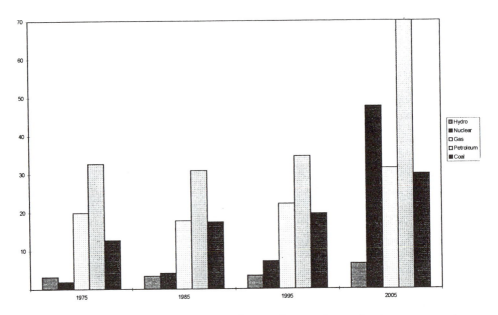

Figure 6.9 Another unsatisfactory, but frequently seen, format for the energy graph.

duce which energy source is predicted to supply less power than coal? If you want to help the reader to understand what is happening, there is no rational reason for using this *3-D, stacked bar chart* to plot time series that you want to compare. The "charm" of the 3-D presentation adds nothing of value. Nor, in fact, is there any value in using a conventional (unstacked) bar chart such as Figure 6.9 to show multiple time series.

Figure 6.10 is the straightforward, informative way to show and compare time series. It is immediately evident which energy source is predicted to supply less power than coal. It is also an easy matter to compare the past and predicted absolute and relative growth patterns of the several energy sources.

When Is a "Soar" a "Creep"?: Inflated Dollars

The Great Currency Mountain Was a Hole in the Ground

"Americans Hold Increasing Amounts in Cash Despite Inflation and Many Other Drawbacks," said the headline in the *Wall Street Journal* in 1979, when the inflation rate was close to double digits. Figure 6.11 was how the *Journal* dramatically showed the "steepening pileup" of currency in circulation dur-

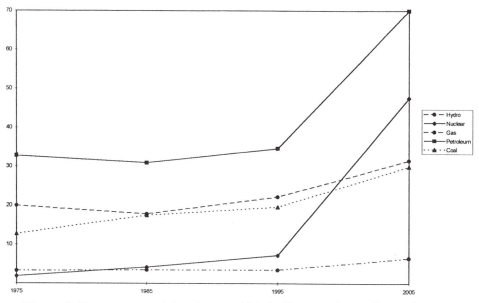

Figure 6.10   Energy graph in a format which facilitates understanding and use.

ing the preceding 25 years: "The amount of currency in individual hands is soaring . . ." (9). The rise, from 1953 to 1979, in the amount of such currency from $28 billion to $102 billion does look like a soar. However, most of the exaggeration of the rise in this graph is due to the failure to correct the value of the currency for its purchasing power.

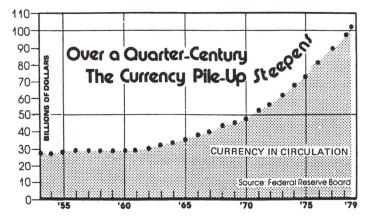

Figure 6.11   Another soar: the piling-up of currency in circulation. (From Ref. 7.)

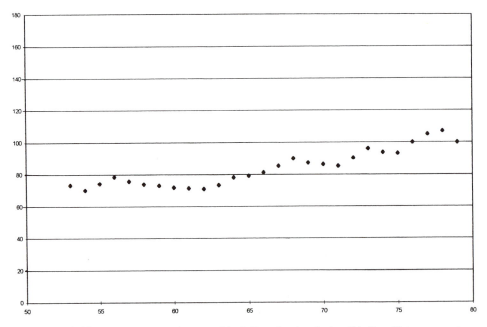

Figure 6.12 Figure 6.11 redrawn with dollars in circulation "deflated" to account for the changes in the Consumer Price Index for the years 1953 to 1979.

What were the real changes in the value of the currency held by Americans? One way of measuring the change in value is to correct the dollar amounts by the change in inflation to get the *purchasing power* of the currency in circulation. We find the purchasing power by reducing the "current" dollar amounts by the Consumer Price Index (CPI) to get "constant" 1979 dollars. These constant dollars are plotted in Figure 6.12 to show the purchasing power of the currency in circulation as a function of time. Clearly, in terms of purchasing power, the currency in circulation is not "soaring" upward; it is "creeping" upward. The average annual rate of growth of the estimated amount of currency in circulation over those 25 years was 1.2%.

What has happened since 1980 to the amount of currency in circulation? Figure 6.13 shows the *deflated* amount of currency in circulation for the years 1953 to 1995. Here, you can easily see that in 1980, the inflation-adjusted amount of currency in circulation really did start to soar. The average annual rate of increase in deflated dollars during the period 1980 to 1995 was 3.8%, more than three times as fast as during the period the *Wall Street Journal* thought there was a "steepening pile-up." Now *that* is a soar!

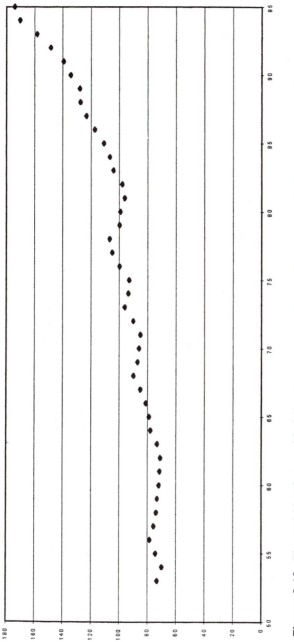

Figure 6.13 Figure 6.11 redrawn with dollars in circulation "deflated" to account for the changes in the Consumer Price Index, updated to cover the years 1953 to 1995.

Chapter 6

MORE ACTION IN THE GRAPHICAL HORROR SHOW

False Proportion: Inflated Dollars

When a Picture Isn't What It Seems to Be

Real wages are wages corrected for the increase (or decrease) in the cost of living due to inflation (or deflation). For example, if the average factory wage rises from $20,000 to $21,000 in the next year (an increase of 5%), we cannot know what happened to *real wages* without knowing the change in the cost of living. If, from one year to the next, the increase in cost of living was 5%, then real wages were not changed. If the cost of living increased by only 2.5%, then real wages *rose* about 2.5%; if the cost of living increased by 7.5%, then real wages *fell* about 2.5%.

The pain of lost value in wages due to inflation is real, but its illustration in Figure 6.14 is false. The *AFL-CIO News* graphically reported the 1978 decline of 3.4% in real wages in Figure 6.14 (10), which the casual reader at

Figure 6.14   Price levels rise by 3.4%, but this illustration shows how it feels, not what it is. (From Ref. 10.)

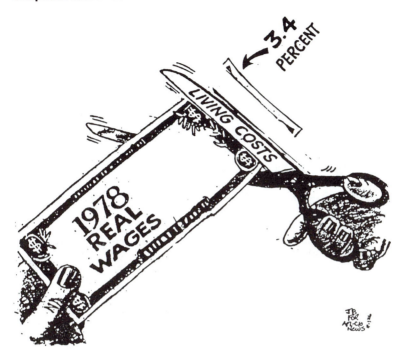

Figure 6.15 Figure 14 redrawn to true scale.

that time would have seen as a severe reduction. The decline was grossly emphasized by showing scissors cutting real wages at a point corresponding to a 40% decrease. This certainly would wake up the reader, but wasn't this a sinful way to do it? This is known as the sin of false proportion. If the graph had illustrated the decline in true proportion (Figure 6.15), the impression would have been substantially different. Stay on your guard, as graphical illustrators are constantly committing this sin.

Visual Confusion: Just Plain Confusing

Now and then, we see a graph that is just plain incomprehensible. Figure 6.16 is supposed to help us understand heat conservation in the home (11). The title claims that it shows us sources of heat, in percentage. If that is all that is shown, why is the bar for the conventional home longer than the other three, which are all nearly, but not quite, the same length. What does the length of the bar mean, if anything? Can it stand for the proportion of conventional fuel? If so, then how can the 75% and 85% bars be the same length? No clue is given. Does this chart help you to understand the distribution of sources of heat?

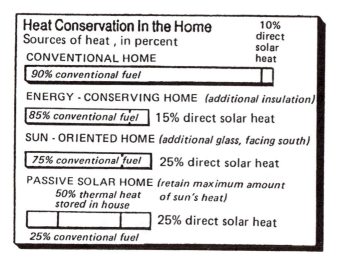

Figure 6.16 How are we to interpret this uniquely confusing plot about heat conservation? (From Ref. 11.)

Note also that the labels are inside three of the bars, at the end of one bar, and below the bottom bar. This is a magnificent example of the sin of visual confusion. It is hard to believe that the originator of this chart understood the distribution of heat sources in the home.

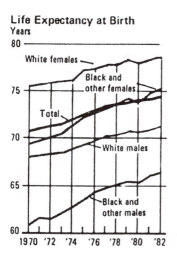

Figure 6.17 Small and confusing. How do we separate "Total" and "Black and Other Females"? (From Ref. 12.)

Small and Confusing Labels

Crowding to the Point of Incomprehensibility

The *Statistical Abstract of the United States* gives us some additional examples of variations on this theme. Figure 6.17 is a plot of Life Expectancy at Birth (12). This graph is confusing for several reasons: There are unnecessarily long and bewildering labels on some of the lines and a visual conflict caused by the bold line ("Total") cutting across one of the others. This is a small graph and the type is small. It needs short labels, which are easy to read. Why do we have to puzzle out the meaning of "Black and other females" to compare with "White females"? How much easier it would be for the interested reader if the graph maker had identified the two lines as "White females" and "Nonwhite females" so they could easily be identified as applying to complementary groups. *Note:* when we searched for a more recent example of this misuse in the *Statistical Abstract*, we could find none. Someone must have read our first edition!

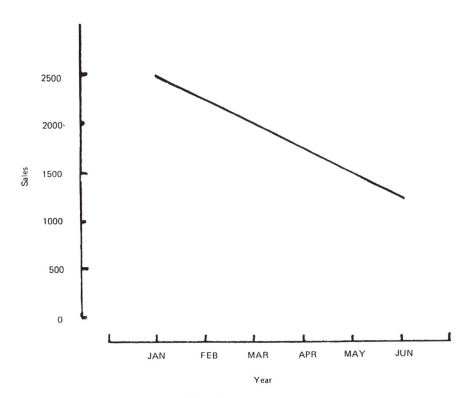

Figure 6.18 Depressing monthly sales.

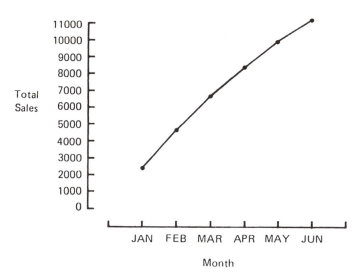

Figure 6.19   Encouraging cumulative sales from the monthly sales of Figure 6.18.

Confusing, Misleading Legends

Some days Dr. Nowall feels completely frustrated. No matter how much he computes and plots he can't get a headline-worthy graph. What to do? If he is to attract the attention he wants, he needs some way to show a "soar" or a "dive"—something that looks exciting or threatening to people who are reading on the run. On one of those black days, he thought up the following procedure reported by Black (13). Other observers have brought us examples of this same misuse, in a number of different contexts.

Suppose that a company experiences declining monthly sales starting from a high of 2500 units in January and declining to 1250 units by June.* The decrease is substantial, 250 units per month, as shown in Figure 6.18. Dr. Nowall consults to this company and is embarrassed by this slump; he sees his continued employment in jeopardy. To keep up management's morale, he plots *cumulative* sales as shown in Figure 6.19. Thus, good cheer can be taken from the report that in January 2500 units were sold and that 11,250 units have been sold by June. Pity the poor viewers of this graph who did not notice the difference between "[monthly] sales" and "total sales."

SUMMARY

Neither Sinned Against Nor a Sinner Be

The rules are so simple! You can easily avoid being a graphical sinner. Similarly, you can easily spot misuses if you know the cardinal sins:

---

*These are contrived data based on a real occurrence.

1. *Missing zero.* The vertical scale should go to zero, or show clearly that it does not, by a break in the axis. If you want to show changes in some data, then plot the changes.

2. *Missing baseline.* If the data are values of an index, show the baseline.

3. *Distorted horizontal axis.* The horizontal axis should be complete, with no breaks and no midstream changes in scale.

4. *Distortion by use of an arithmetic vertical scale for economic time series.* Any series of data growing at a constant percentage rate per year will "soar" when plotted against an arithmetic vertical scale. Replot any such data on a logarithmic scale. If you are plotting data for an economic time series, or you have other reasons to be interested in percentage changes, then you also should use a logarithmic scale on the vertical axis.

5. *Inflated dollars.* If the data you are plotting are monetary values plotted in current (actual) values, then replot them in deflated (or constant) values to determine their value in purchasing power, and label appropriately.

6. *False proportion.* If the purpose of a picture of an object is to show a relative amount, check that the illustrated relative proportion is valid.

7. *Visual confusion.* Don't make confused or confusing graphs, and refuse to accept them in others' work.

8. *Confusing labels.* Don't make confused or confusing labels, and refuse to accept them from others. Apply the standards of clear communication and correctness to the labels of a graph.

9. *Crowded to the point of incomprehensibility.* Don't do it, and don't accept it.

10. *Great differences in the lengths of the axes.* Again, don't do it, and don't accept it. There are many informed opinions and an immense literature on the perception of form. Fred H. Knubel, director of Public Information for Columbia University, basing his opinion on many years of experience presenting information in graphical form, has concluded that the least biased presentation is an approximation to a square. Edward R. Tufte concludes his book on graphics with this statement, based on his study of the literature: "If the nature of the data suggests the shape of the graphic, follow that suggestion. Otherwise, move toward horizontal graphics about 50 percent wider than tall" (2, p. 190).

Remember the authors' definition of a misuse: Using numbers in such a manner that—either by intent, or through ignorance or carelessness—the conclusions are unjustified or incorrect. Scrutinize your graphs to reduce the likelihood of you or a reader drawing an unjustified or incorrect conclusion.

7
Methodology: A Brief Review

A workman is known by his tools.
—Anonymous proverb

The king lost his way in a jungle and was required to spend the night in a tree. The next day he told some fellow traveler that the total number of leaves on the tree were "so many" (an actual number was stated). On being challenged as to whether he counted all the leaves he replied, "No, but I counted the leaves on a few branches of the tree and I know the science of die throwing."
—Ancient Indian epic Mahabharat (Nala-Damayanti Akhyān)

INTRODUCTION

There are many hundreds of useful tools—statistical methods—for analyzing data and drawing conclusions. At the most basic level, we use simple, straightforward tools for which pencil-and-paper arithmetic is adequate. At the other end of the spectrum, we use complex, sophisticated procedures that can require the fastest available computing systems for timely results.

Like all tools, the effectiveness of the statistical methods depends on using them appropriately. Use a crowbar to change the tire on a mountain bike, and you risk damaging the tire or the wheel and finishing your trip on foot.

The analogy with the mountain bike tire is a valid one. The purpose of a crowbar is to provide leverage for heavy tasks, such as moving boulders. Changing a bike tire calls for a lightweight tire iron which provides the appropriate leverage. The creative individuals who develop statistical methods to analyze data must make assumptions about both the user-analyst's purpose and the nature of the data, so that the tools they devise are suitable to these constraints. You may need only a pocket calculator or have to use a computer capable of beating the world chess champion to solve your problem. Making that choice is the most creative element in the project, and it can make or break your analysis. For example, the statistical analyst's purposes may be to:

> Make inferences about the nature of situations or populations which must be sampled
> Evaluate the effect or effectiveness of some approach, method, treatment, compound, program, and so forth
> Compare two or more approaches, methods, and so forth
> Explore the data, looking for suggestions of future directions
> Summarize the data to make them more comprehensible

The analyst may know or can assume that the data:

> Come from a population having a particular distribution of values
> Have no measurement error
> Come from a situation in which the causes of variation are unchanging
> Are randomized to meet a need of the situation
> Have a known variability
> Have the same variability as other data in a comparison, or do not change variability as some factor of interest changes
> Are from a random sample
> Contain biases which can be corrected

Thus, the analyst's purposes and the attributes of the data determine the statistical tool to be used.

STATISTICAL TOOLS

Even such basic arithmetic tasks as counting, adding, subtracting, taking percentages, ranking in order, making plots, and so forth, are tools that help us to understand the messages hidden in the data, and communicate them to others. More complex tools are often concerned with *summarizing* data so that we can draw some conclusions without looking at the data in detail. Often we can learn more from the summaries than we can by working with the actual data. Examples of such tools of summarization are mean, median, standard deviation (a measure of the scatter, or dispersion, of the data), correlation coefficient, and coefficients in regression equations. In experiments, the tools of statistical analysis deal with the design and analysis of the complex and often large and extended experiments by which we obtain the data. When an analyst selects the wrong tool, this is a misuse which usually leads to invalid conclusions.

Incorrect use of even a tool as simple as the mean can lead to serious misuses. For that reason, we start with a preliminary discussion of some of the simple and most used summary tools of statistics, and use them to illustrate how misuses can occur. More complex tools, such as time series analysis, design and analysis of experiments, discriminant analysis, and analysis of variance are extensively discussed in statistical texts and scholarly literature.

But all statisticians know that more complex tools do not guarantee an analysis free of misuses. Vigilance is required on every statistical level.

On the Average

The most commonly used statistical summary measure is a "typical value" for a set of data. Why would someone want a typical value for a set of data? The reasons are many. An athlete might want to know the typical time for a particular knee injury to heal. An engineer might want to have a typical value of the strength of a structural member. A regulatory agency might want to know the typical time for a pollutant to disappear from soil. An investor might want to know the typical annual return of mutual funds in an industry sector.

Because statisticians usually think of the numerical values of data for a single variable as falling along a line, they often call a typical value a measure of "location." When viewed this way, the typical value tells where the numerical values fall along some imaginary axis. In the case of the athlete, the values could be anywhere along an imaginary axis from zero to infinity. But what is the "location" of the values reported? Somewhere between 50 and 100 days? Some people will heal in 5 days, others may have to wait as long as 300 days. Thus, the concept of location.

Statisticians may also think of a typical value as a measure of "central tendency," showing where the data tend to cluster. For example, if the data were mostly near 75, and roughly symmetric above and below 75, the central tendency would be 75.

The data in Table 7.1 are the annual salaries of 10 business executives. Below, we look at two of the statistical tools (which most statisticians would call "measures") that give us a statistical typical value.

The Mean

The *arithmetic mean,* usually called "the mean" or "the average,"* is the sum of all data values divided by the number of such values. Thus, for the executive

---

*Technically, this is a misnomer. For example, the U.S. Census Bureau says, "An *average* is a number or value that is used to represent the 'typical value' of a group of numbers" [1, p. xvii].

Table 7.1 Annual Salaries of 10 Executives
(in thousands of dollars)

Raw data
890
1,110
1,460
1,420
2,000
1,430
1,520
1,110
2,400
1,680

salaries in Table 7.1, you compute the arithmetic mean by adding up all the salaries and dividing by the number of executives. In this case, the total for all the salaries is $15 million; divided by 10 you get a mean executive salary of $1.5 million. This is the value of salary that, if all executives in the study made the same salary, would give the same value as the sum of all their incomes.

The arithmetic mean has the most meaning when the values are closely centered, with few exceptional values and tending to symmetry about the mean. But suppose that the one executive who earned $1,460,000 has had a profit-sharing bonanza one year and earned $5 million more for a total salary of $6,460,000 instead of $1,460,000. While most of the executive salaries are still "around" $1.5 million and only one other makes more than $2 million, the mean has jumped from $1.5 million to $2 million, an increase in the value of the mean of more than 30%. Clearly, the mean is not a good indicator of "typical values" in this type of situation.

Similar situations often arise wherever money or value is involved. For instance, the appraised values of almost all of the homes in an enclosed neighborhood may lie between $300,000 and $500,000; but one home is appraised at $2 million. The arithmetic mean would include that extraordinary home in the computation, and give a misleading "typical" value for the homes in that neighborhood.

The Median

Using the next most common measure of central tendency, the *median*, avoids this problem. To get the value of the median, take all the numbers you have collected, and order them by increasing value. Once the numbers have been ordered, the median is the middle value (if the number of values is odd) or the

Table 7.2 Annual Salaries of Ten Executives,
Arranged in Order of Increasing Magnitude
(in thousands of dollars)

890
1,110
1,110
1,420
1,430
1,460
1,520
1,680
2,000
2,400

average of the two middle values (if the number of values is even). To get the median of the salaries in Table 7.1, order the values as shown in Table 7.2. Then find the middle value (or as in this case, the average of the middle two values) to get a median executive salary of $1,445,000 ($1,430,000 + $1,460,000 divided by 2).

Thus, the median is not much affected by widely dispersed values. For this reason it is often used for reporting typical salary, age, and similar values. The median is also useful because it is easily understood: it is that value that about half the population have values below and half have values above. In 1994, the median money income of the U.S. households was about $32,300; about half of the 99 million households had money incomes below $32,300 and about half had money incomes above $32,300 (1, Table 709).

Note that for the original data set of Figure 7.1, the median of $1,445,000 is only a little less than the arithmetic mean $1.5 million. But when the lucky executive's $1,460,000 salary is increased to $6,460,000, the median does not change. At $1,445,000, the median is still typical of the executive salaries. The mean does, however, and the new mean of $2 million is not typical.

Clearly, the mean is greatly influenced by extreme values, but it can be appropriate for many situations where extreme values do not arise. To avoid misuse, it is essential to know which summary measure best reflects the data and to use it carefully. Understanding the situation is necessary for making the right choice. Know the subject!

Managed Care for Measures of Typical Values

In 1997, Columbia/HCA, then the world's largest health insurer, came under fire for its charging policies. In an article on Columbia's prices in *The New*

York Times, Martin Gottlieb and Kurt Eichenwald reported that for 1995 the median charge for a Columbia hospital stay was $4370, 8% higher than the median for other hospitals (2). According to the *New York Times* article, a Columbia/HCA vice president disputed the findings, taking issue with the way Ohio University's Center for the Health Care Industry Performance Studies analyzed the company's price data:

> Of the available measures, you have selected the one that puts Columbia in the least favorable light. . . . Columbia . . . chose not to analyze [the data] in the manner used by those researchers [Ohio State]... and by many consulting organizations. Instead, the company used three other methods based on averages other than medians. Using two of those, Columbia found its prices to be roughly 5% higher than the norm. Only one method showed Columbia with better prices, at 7% below the average for all other hospitals. But that method did not exclude hospitals with the most extreme high and low prices, as is generally done to avoid skewing the outcomes [2].

Note that while the median does not *remove* extreme values, it does reduce their effect.

Different summary measures of what is typical give different results. That is all we propose to show in this example. However, to determine whether Columbia's charges are out of line goes beyond summary measures and requires an in-depth audit of their pricing history and policies. Subsequent to the statements we report above, governmental auditors carried out such audits. The result was a major reorganization of Columbia/HCA and the departure of several executives.

Sometimes It's Hard to Count Right

As we mention in the introduction to this chapter, even simple arithmetic is a tool, a method of statistics. And it is sometimes hard to count right, as you shall see. True, our first example involves counting with a computer. In general, the computer just makes counting easier and more accurate. But not always.

The United States has experienced periods of both prosperity and mild recession since the 1970s. Despite the down periods and problems such as a growing disparity in income distribution, the U.S. economy has shown remarkable resilience and capacity to grow. Given that long-term record, it is hard to visualize the types of arguments put forward in the '70s about the negative effect on the U.S. economy due to "entitlement" programs such as Social Security.

However, in 1974, Dr. Martin S. Feldstein, who subsequently served as chair of President Reagan's Council of Economic Advisors, announced the results of a massive data analysis that showed that Social Security deductions reduced personal savings by 50% and made the country's plant and equipment 38% smaller than it would have been without Social Security. Then, in 1980, "Professor Feldstein acknowledged [a] mammoth mistake ... [which] led to a multi-billion dollar overestimate of the negative effect of Social Security on national saving" (3).

Do economists often make mistakes of such magnitude? Alan Blinder, an economist, says, "There are probably untold numbers of errors buried in the economic literature. . . . If you had made a small programming mistake ... it would probably not be discovered unless it had produced crazy numbers, and Marty's [Feldstein] did not" (4). In fact, "Marty's" calculations did produce results which, if not "crazy," were strange enough to arouse the suspicions of two other economists, Dean Leimer and Selig Lesnoy. They decided to check Feldstein's results, since they knew that "other major studies have yielded smaller estimated effects on saving or concluded that there was no evidence of a significant effect" (5).

Why wasn't Professor Feldstein more suspicious? We can't say for sure, but perhaps it was his strong belief in his conclusions, even if erroneously obtained, for after the "mammoth" error was found, he stated: "My sense is that there is a general belief in the profession that Social Security still depresses savings, although the evidence is not finally in on the magnitude of that effect" (3). He has also argued that: "When [a] legislative change is taken into account (and the error in the Social Security wealth series is corrected) the results are very similar to the conclusions reported in my earlier study" (6). The disagreement among experts and the results of analysis so far lead to only one reasonable conclusion: that we did not and still do not know with any precision the effect of Social Security on savings. But the strongest and most widely publicized conclusion on this issue was based on an analysis with a "mammoth" computational error. Interestingly enough, Feldstein's argument about this particular negative effect of Social Security on savings has rarely been raised again since the early 1980s. Perhaps Feldstein's dire prediction rooted in error was forgotten in the light of the generally good performance of the U.S. economy in the boom years of the late 1990s.

Correlation

Don't Move the Kids, Move the State Capitols

North Americans were debating about the best way to educate their kids well before 1776 (as we mention in Chapter 2), and they have been at it ever since.

Arguing at a Democratic Party gathering against the idea that school expenditures and pupil achievement are related, Senator Daniel Patrick Moynihan claimed (facetiously, of course) that, "if you would improve your state's math scores, move your state closer to the Canadian border!" (7). To demonstrate the use and misuse of several tools, we evaluate, using current data, Senator Moynihan's provocative statistical claim.

First, an introduction to the relevant methodology. The *scatterplot* is a plot of two numerical variables which can be assigned to a series of points plotted at positions corresponding to the value of one variable on the horizontal axis and the value of the other on the vertical axis. The senator's primary variable of interest is the sample average national score by state on the Grade 8 Public Schools mathematics examination, conducted by the National Center for Education Statistics in 1996 (8). It is no surprise that the average statewide scores vary from state to state. To "explain" the variations in average state mathematics test scores, Senator Moynihan used the distance of each state's capitol from the Canadian border.

We make our scatterplot of average mathematics scores versus the distance of the state capitol from the Canadian border by plotting the score along the scatterplot's vertical axis and the distance from Canada along the scatterplot's horizontal axis as shown in Figure 7.1. Each point in Figure 7.1 corresponds to a given United States state; its position on the scatterplot is determined by the values of the two variables (distance, score) for that state.*

You can see from Figure 7.1 that there is a negative relationship between the test scores and the distance from Canada. By "negative," we mean that as the distance from Canada increases, the mathematics score tends to decrease. We used the term "tend" because the points do not lie on a straight line of direct proportionality.

The strength of such an association is measured by the *correlation coefficient,* another summary tool. It is easy to compute this coefficient with a computer or a calculator. The correlation coefficient ranges from −1 through 0 to +1. As in the case of measures of typical values discussed at the beginning of this chapter, there are a number of different ways to compute a correlation coefficient. Alas, each may give a different value for the same data. The "Pearson product-moment correlation coefficient" is most commonly used and is denoted by "R" or "r." This is the correlation coefficient you can compute using a pocket calculator.

A value of $R = +1$ is an association such that points lie on an upward-going line (as shown in Figure 7.2), and $R = −1$ to a similarly tight association

---

*There are fewer than 50 points, since not all states satisfied the guidelines for inclusion. We also excluded Hawaii and Alaska, because it would be hard to move them closer to the Canadian border, and Washington DC, which is not a state with a capital.

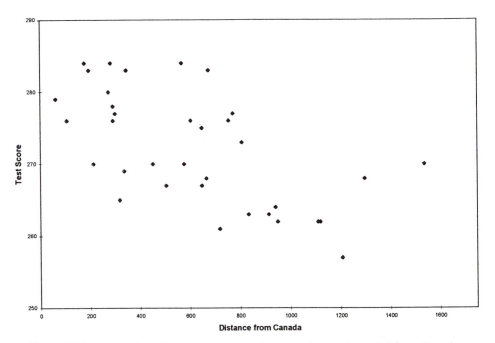

Figure 7.1 Scatterplot of test scores versus distance of the state capital from Canada.

on a downward-going line (as shown in Figure 7.3). Examples of intermediate values of R are shown in Figures 7.4 and 7.5. Figure 7.5 shows a "loose" downward-going association between two variables.

We get a value for the correlation coefficient of $R = -.61$ from these data. The scatterplot and the value of R show that the average math score and distance from Canada for states are *associated*. That is, they are paired together in a decreasing (negative) relationship.

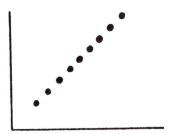

Figure 7.2 Scatterplot showing perfect positive correlation ($R = 1.0$)

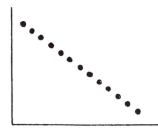

Figure 7.3 Scatterplot showing perfect negative correlation ($R = 1.0$)

Someone like Dr. Nowall would believe that Figure 7.1 supports the contention that there is a cause-and-effect relationship between distance to the Canadian border and math scores. But to conclude that closeness to Canada *causes* an increase in scores is a leap from association to *causation*, a common misuse. Senator Moynihan, with his customary twinkle, suggests that a policy of moving states closer to Canada is the key to improving educational achievement, mocking some of the improper leaps from association to causation.

To take the relationship between distance to Canada and test scores seriously would be a misuse of statistics by *confusing association with causality*. Although the suggestion is clearly absurd in this case, such reasoning is a common source of false generalizations about the behavior of people, organizations, countries, ethnic groups, and many phenomena in the physical world. John Tukey, one of the great statisticians of our time, says that to support the notion of "cause," we need *consistency, responsiveness,* and a *mechanism* (9, p. 261). To show just how badly Dr. Nowall (and many of Senator Moynihan's colleagues, some of the media, and statistically challenged researchers) has misused statistics, we give a short discussion—far from complete—on cause.

Figure 7.4 Scatterplot showing less than perfect positive correlation (R is between 0 and 1)

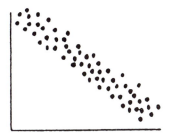

Figure 7.5 Scatterplot showing less than perfect negative correlation (R is between 0 and 1)

Let us imagine the following situation. The community of Hypertown finds that its drinking water contains a certain amount of a pollutant. Also, its inhabitants have, on the average, "too high" blood pressure. The health commissioner of Hypertown looks at surrounding communities and finds another community, Hypotown, which has none of this pollutant in its water supply. The average blood pressure of its inhabitants is lower than that of Hypertown residents. If this pattern is found in a number of other pairs of communities, then we have *consistency* of findings.

To confirm *responsiveness,* you must show that changing the level of pollutant in the water supply changes the average blood pressure, at least up to a point. There is no necessary upper limit to how much pollutant there may be in the drinking water unless the "water" consists solely of pollutants, in which case it is no longer water. But there is certainly an upper limit to blood pressure for specific individuals. If it gets too high, the person dies.

In general, to confirm responsiveness you must run an experiment. In this case, you would like to have an experiment in which the amount of the pollutant fed to human subjects was varied and their blood pressure was monitored to confirm or disprove that the rise and fall of blood pressure is directly related to the amount of the pollutant consumed.

To support the existence of a *mechanism,* you must be able to construct a model of the real world based on known principles which can, perhaps step by step, explain the effect. In the example we are considering, such a mechanism could involve showing that the pollutant directly acts on cells in the kidneys to affect the retention of sodium and water in ways that have been shown through medical research to produce elevated blood pressure.

To prove a *causal link* you must show that the effect that you have observed does not produce the presumed cause. In our example of the pollutant and high blood pressure, if we did not have a clearly demonstrable mechanism, we would also have to show that it was not something about people with high blood pressure that makes them drink water containing the pollutant.

Thus, if you have no more than a statistical demonstration of association between two variables (such as distance from Canada and math scores), you have no basis for saying that one causes the other. You do have a basis for suspecting that this might be so and then working to confirm or disprove your suspicion.

As for improving student achievement by moving states closer to Canada, before uprooting millions of people and institutions, we would have to show through consistency, responsiveness, and a mechanism that moving toward Canada would improve the mathematics scores.

There may be no causal mechanism at work here, and we have good reason to believe that moving a state would not *cause* an increase in student achievement. On the other hand, there are many possible reasons that distance from the Canadian border would be *associated* with lower test scores, which we discuss in the next section.

Simple Regression

Will States Undertake a Major Move Northward to Improve Educational Achievement?

Linear regression is a complex statistical tool. Even for data sets of modest size, the computations are tedious to perform with pencil and paper. But we can now carry out these computations with computers and calculators. The tedium has been eliminated but difficulties in interpreting the results remain, as we show by continuing our fictitious, but not unrealistic, example.

A naïve person, such as Dr. Nowall, who is not familiar with the concepts of causation may continue to pursue Senator Moynihan's proposal to increase educational achievement by selectively moving states to the North. By how many kilometers does a state have to move to increase the scores by a given amount? If one state moving to the North displaces another to the South, by how much will the states change their scores?

If we draw some "best" straight line through the data shown in the scatterplot, we can answer these and similar questions by using the straight line to extrapolate values of math test scores for a given distance from Canada. You can easily do this by computing the *least-squares regression line* on a computer or pocket calculator. These tools will compute the two parameters A and B for the formula ($Y = A + B * X$) for the straight line through the data that gives the least value for the sum of the squares of the "residuals" (vertical distances between the line drawn and the plotted points).*

---

*And is, for that reason, sometimes known as the "least squares line."

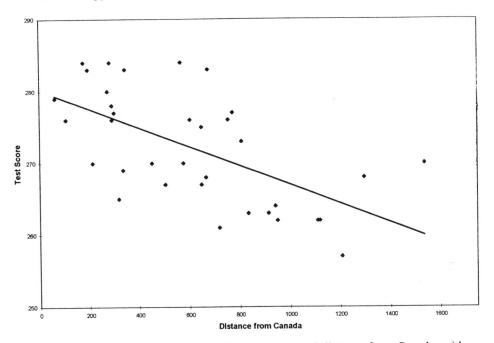

Figure 7.6 Scatterplot of the data for test scores and distance from Canada, with least-squares regression line.

Enter our mathematics test score data into a linear regression computer program, and you will get the formula $Y = 280 - 0.013 * X$, where Y is the average math test score and X is the distance from the Canadian border in km. This line is shown in Figure 7.6. You can see that the regression line has a "Y intercept" of 280; this is the score for $X = 0$, a state whose capital is 0 km from the Canadian border. The "slope" of the line is –0.013; for every km farther from the Canadian border, the score drops by 0.013 points. A thousand kilometers is worth 13 points! Move 2000 km, and a state can move from a nearly minimal mean score of 260 to a mean score of 280, close to the top. Have we proved that distance from the border causes low scores, or only that this is a relationship of association? Once again, to prove cause and effect, we must show consistency, cause, and mechanism. Can we prove this, when there are several other variables which may be causal factors? These could include class size, economic factors, teacher training, and so forth. The jury is still out and may stay out for a long, long time.

You Won't Surprise Us if You Go Bankrupt

People want to know what the future holds, but even a statistical crystal ball is not always reliable. *Extrapolation (forecasting)* is a statistical tool which has been of considerable value, particularly in science and engineering. But it can also be treacherous, especially in the hands of the reckless, the unskilled, and the sensation-seeking.

Many methods (such as linear regression) are used to project future values, and in certain circumstances (control of inventories, early warning of cost overruns, and so forth), one or another of these methods can be useful, but all have problems of accuracy because no method can reveal the "truth" about the future. Careful prognosticators refer to "projections" and not "predictions." Only headline seekers label their guesses as "predictions." Here is an example from daily economic life in the United States.

Personal bankruptcy is a means by which an individual can financially "wipe the slate clean" and get a fresh start with zero indebtedness. However, a personal bankruptcy often seriously affects the person's ability to get credit. For that reason, personal bankruptcy is a measure that should be taken only in extreme situations, and most financial advisers recommend against it.

Nonetheless, personal bankruptcies in the United States are on the rise. The number of filings for personal bankruptcy has increased by about 2.5 times in the 10 years from the second quarter of 1986 to the second quarter of 1996 (10). Banks have legitimate reasons for keeping confidential their losses on credit card accounts, but we imagine that the amount due to personal bankruptcies must be significant.

Because of his own recent problems managing his credit card indebtedness, Dr. Nowall hypothesized that there might be a causal relationship between the amount of outstanding credit card balances and personal bankruptcy. If he could show this relationship in a regression formula, this could help banks to protect their interests through defensive policies. Then, as a newly qualified bank consultant, he might be able to more rapidly pay off his own credit card indebtedness.

Dr. Nowall obtained the quarterly number of nonbusiness bankruptcies for the years 1985 to 1996 (10) and the total amount of U.S. outstanding credit card balances (11), as shown in Table 7.3. He called up the linear regression program on his personal computer and let the computer find the linear relationship between the bankruptcy filings and the outstanding credit card balances. The equation of this line is $Y = 78 + .62 * X$, where Y is the number of personal bankruptcy filings in thousands and X is the outstanding credit card balance in billions of dollars. Dr. Nowall draws several conclusions from this equation. First, he feels that he now knows the *rate* at which bankruptcies increase with balances; for every billion-dollar increase in the outstanding balances, the number of personal bankruptcies increases by 620 (0.62 thousand). Secondly, he can estimate the number of bankruptcies if he has a value for the expected balances. If the balances are $200 billion, then he estimates the number of bankruptcies as 202,000 (0.62 times 124 plus 78).

Table 7.3 Total Outstanding Balances and Personal Bankruptcies by Year and Quarter

Year	Quarter	Total Outstanding Balances	Personal Bankruptcies
85	1	55.6	73
	2	59.1	84
	3	62.7	88
	4	69.6	96
86	1	69.9	103
	2	71.6	114
	3	74.6	115
	4	80.7	116
87	1	78.5	119
	2	81	123
	3	86.1	124
	4	95.5	127
88	1	94.7	134
	2	98.3	138
	3	103.3	139
	4	113.4	139
89	1	112	145
	2	116.2	158
	3	121.6	153
	4	133.2	161
90	1	131	156
	2	136.3	180
	3	141.6	177
	4	153.8	194
91	1	148.8	213
	2	152.1	228
	3	156.2	214
	4	166.4	217
92	1	159.2	234
	2	162.7	233
	3	167	220
	4	178.8	212
93	1	173.6	206
	2	180.5	213
	3	189	203
	4	206.3	193
94	1	204	193
	2	216.6	203
	3	231.8	208
	4	257	202
95	1	260	213
	2	279	235
	3	295	234
	4	321	245
96	1	321	266
	2	335	297

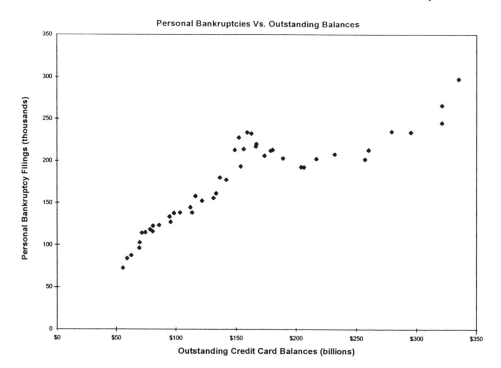

Figure 7.7 Scatterplot of personal bankruptcies vs. outstanding credit balances.

He computes the correlation coefficient to be 0.89, which he thinks shows that the relationship between balances and bankruptcies is quite strong. All of this reinforces his opinion that personal bankruptcy is *caused* by consumers running up excessively high balances on their credit cards. If a bank works to reduce the amount of outstanding credit card balances, it will reduce the number of personal bankruptcies of its card holders. He is now in a position to sell his services as a financial analyst to the banks.

This example, which is based on a bank officer's actual proposal, is a classic example of a misuse of linear regression to predict an outcome. First, Dr. Nowall failed to look at the scatterplot, which we show in Figure 7.7. Had he done so, he would have seen that there are three separate regions that may result from changes in the causal processes by which the association between the variables is generated. In the first region, from values of outstanding balances from about $50 billion to $150 billion, the number of personal bankruptcy filings rises at a rapid rate. In the middle region, from about $150 billion to $250 billion, the number of filings actually *decreases* as the amount of outstanding balances rises! Then, for values of outstanding balances above $250 billion, the filings rise at a rate similar to the rate for less than $150

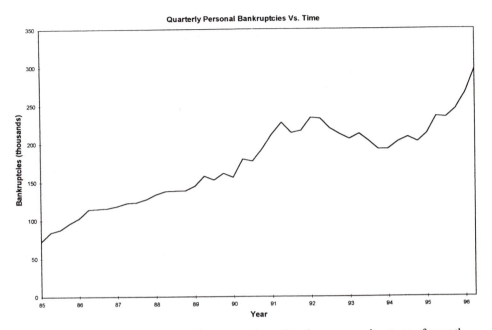

Figure 7.8 Personal bankruptcies versus time, showing a general pattern of growth in three "phases."

billion. It is a first principle of regression analysis with only two variables (as in this case) or three variables (which can also be plotted), that the investigator *look* at the graphical representation of the data. This is not the only way to discover such relationships, but where it can be done, it is the easiest way.

When two variables are derived from time series, as are these two variables, it is easy to obtain a "nonsense correlation" from the causal effects of the common variable, time. If you scatterplot beer consumption and the number of teachers in the U.S., you will find a high positive correlation. Is this because teachers are high consumers of beer? Not at all. It is simply that, with the growth in population, both of these series are increasing with time, so that high values of one variable are associated with high values of the other.

Look at the time series for personal bankruptcy filings and outstanding credit card balances, shown in Figures 7.8 and 7.9. Both series are increasing with time, so a positive relationship with a high correlation is hardly a surprise. Compare the pattern of the bankruptcy filings versus outstanding balances of Figure 7.7 and Figure 7.8; they are almost identical. Of course! The outstanding balances (Figure 7.9) are increasing steadily with time, so the scatterplot of Figure 7.7 is essentially the time series plot of Figure 7.8.

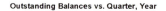

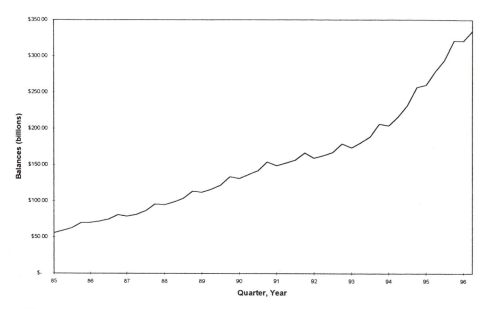

Figure 7.9 Outstanding balances versus time. Note the steady upward progression, as opposed to the growth of personal bankruptcies in Figure 7.8 with a clear break-down into three phases.

Perhaps an increase in personal bankruptcies is caused by the increase in outstanding credit card balances. But if this is so, how do we account for the negative relationship for balances of $150 billion to $250 billion? We can see from Figure 7.8 that this negative relationship results from the decline of bankruptcy filings during the years 1991 to 1995.

Some causes are at work, but it is not a simple relationship between bank balances and bankruptcy filings. As we noted in Chapter 3, Know the Subject, there is no substitute for knowledge of the subject matter. While we make no claims to expertise in this area, we can imagine some of the other variables, economic and social, that might be causal influences on personal bankruptcy filings. These could include the rate of divorce, attitudes toward bankruptcy, unemployment levels, interest rates, installment debt level, credit policies of card issuers, and many others.

Predicting Who Will Be a Good Student

We administer tests to millions of students every year so that we can decide whether to admit them to colleges and universities. In general, the purpose of

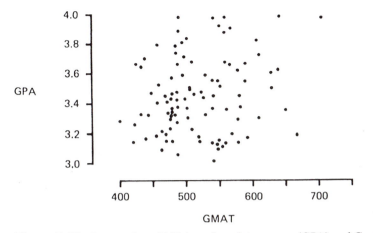

Figure 7.10 Scatterplot of MBA grade point average (GPA) and Graduate Management Admission Test (GMAT) score for 100 students in an MBA program.

admissions policies is to accept candidates with a high probability of success at the admitting institution. The purpose of using scores on standardized tests such as the Scholastic Achievement Test (SAT), Graduate Management Admissions Test (GMAT), Graduate Records Examination (GRE), and so forth, as one of the criterian for admission is to help admissions committees predict probability of success.

In statistical terms, the goal is to use the test score to predict the grade point average (GPA) the student will earn in the admitting institution. Any institution (or group of institutions) can get data on test scores, and the GPA statistics come from the records of currently enrolled students.

One set of such data is illustrated in Figure 7.10, which shows the scatterplot of GPA versus GMAT* for 100 students enrolled in a particular master of business administration (MBA) program. The correlation coefficient for these data is $R = .21$, which drops dramatically to $R = .16$ if the student in the upper-right corner is eliminated from the computation. This one exceptional student has a large effect on the perceived association. This particular institution uses the GMAT as the major component of its decision on whether to admit a student! Would you call this a misuse of statistics?

Before making a judgment, there is one important fact to consider: Since the GMAT was used as a criterion for admission, no applicant with a GMAT score below 400 was admitted. Thus, we have no way of knowing what GPAs

*One of the criteria for admission used by many business schools.

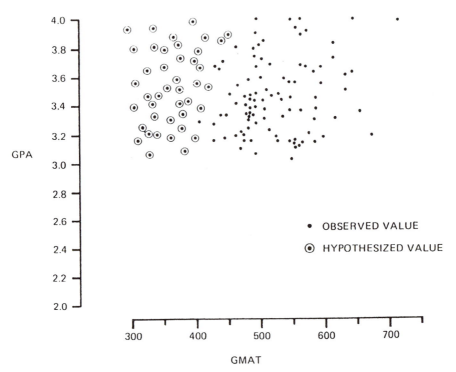

Figure 7.11 Hypothesized scatterplots of student performance versus admission test score are shown in this and the following three figures. The points above a GMAT score of 400 are observations from Figure 7.8. The GPA values corresponding to GMAT scores below 400 are hypothetical values that might have been earned by the students who were not admitted because their GMAT scores were below 400. Which of these scatterplots do you feel is most likely?

individuals scoring below 400 on the GMAT would have gotten if they were students. It is possible that if we had these data we would find there was a much higher correlation between the GPA after admission and the GMAT score before admission. Or, on the other hand, perhaps not. Four possible distributions (including the students below the cutoff value) are shown in Figures 7.11 to 7.14. Which do you think is most likely? The institution's use of these tests is based on the belief that the downward trending distributions of Figures 7.12 and 7.13 are the likely distributions.

It is a misuse of statistics to conclude that a student's GMAT score is a good predictor of the student's GPA *based on only the data of Figure 7.10*. You can draw that conclusion only if you know that the *whole* distribution (including students with scores below 400) is similar to Figure 7.12 or 7.13.

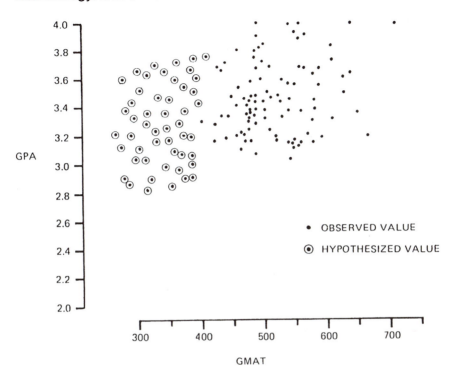

Figure 7.12 A hypothesized scatterplot of student performance versus admission test score.

Nor can you conclude that it is a *bad* predictor in the absence of knowledge that the whole distribution is similar to Figures 7.11 and 7.14. This is a statistical Catch-22. Schools admit students based on the assumption that the test is a good predictor of future academic performance. They do not admit students below a certain test score, and therefore do not get statistical data which would enable them to ascertain whether their assumption is correct.

The Educational Testing Service (ETS), which designs and administers the GMAT, summarized the results of studies at 20 business schools and found that the median correlation coefficient between GMAT scores and first year GPA was .35; values as low as −.12 and as high as +.76 were observed (12). Unfortunately, we do not know the cutoff score for admission for the individual schools. However, the median value of .35 is some support for a belief that *in general* (as opposed to a particular case, such as the school illustrated in Figure 7.10) a higher test score predicts a higher mean GPA. Is there a way to test the correlation more reliably?

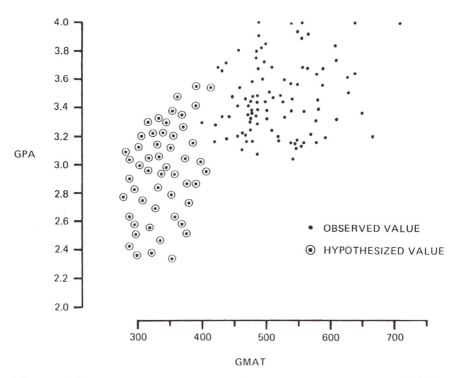

Figure 7.13 A hypothesized scatterplot of student performance versus admission test score.

One way for a particular school to estimate the nature of the distribution below the cutoff score is to admit a small sample of randomly chosen students whose scores are below the cutoff score and track their performance. Of course, the same method could be used for a sample of schools to evaluate the validity of the test more generally.

In today's litigious environment, can you imagine the uproar if a school chose to admit some students with test scores below the normal cutoff point? It is unlikely that the protestors would be impressed by the scientific purpose of such an action. And yet, their whole rationale, that test scores are a valid standard for admission, may (or may not) be based on false premises. Until schools conduct the experiment, the assumption is just that—an assumption.

Another approach would be to compare the observed correlations between schools that have high cutoff values for test scores (which, because of the cutoff of so much of the lower end, have low correlations) and schools that have low cutoff values (which will have a wider range of test scores and would show higher correlations if Figures 7.12 and 7.13 are the underlying distributions).

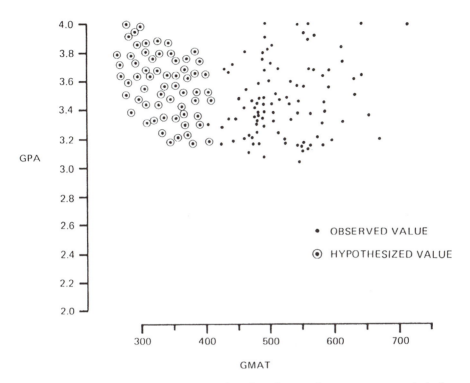

Figure 7.14 A hypothesized scatterplot of student performance versus admission test score.

Multiple Regression

Giving Meaning to the Correlation Coefficient

We now have enough background to return to our consideration of the correlation coefficient and give it some meaning. To some people, a correlation coefficient of $R = .2$ (the particular instance of the school illustrated in Figure 7.10) or $R = .35$ (the median value for 20 schools given in the ETS report) may sound significant in predicting one variable (such as GPA) from another (such as GMAT score). Fortunately, we can give tangible meaning to the correlation coefficient.

We use a simple fictional example to show how to give meaning to the correlation coefficient in some circumstances. The Farr Blungit Corporation sells blungits through 10 retail stores. The sales of blungits, in thousands, during the past month for each of the 10 stores are shown in Table 7.4.

Farr plans to sell blungits through additional stores. *Using only these data* for the 10 stores, what is a reasonable estimate for the monthly sales of

Table 7.4 Last Year's Sales of Blungits at 10 Stores
(in thousands of units)

3.2
7.9
1.8
12.6
12.1
0.8
5.8
10.6
16.0
4.9

blungits in a new store? In the absence of any other information about the store (such as location, the types of potential customers, advertising and sales promotion activities, seasonality, and so forth), you can start by using the average (mean) of these data—7570 units—for your first approximation. But how good is this estimate? The difference between this average and the observed values is quite large, ranging from –6770 to +8430 units!

The measure we have introduced for "spread," or variability, is the *standard deviation*, which in this case is about 5100 units. Statisticians and researchers using statistics also measure the dispersion with the square of the standard deviation, called the *variance*.* In this case, the variance of the blungit sales data is the square of 5100 units, or about 26 million. To predict the sales of blungits in a new store so that you can plan manufacturing, distribution, and staffing of the store, you want to make the best estimate you can from the data. To put the concept of "best" into numerical terms, a statistician might say that Farr should reduce the spread, or variance, from 26 million to a lower value.

How might we reduce this variance? Is it possible that some variable such as those mentioned earlier (type of potential customers, advertising and sales promotion, and so forth) would "explain" some of this variance and give us an estimate of annual sales of blungits with a smaller variance? Such a variable, whether it does or does not reduce the variance, is called an *explanatory variable*. Farr's marketing manager held a meeting to discuss this issue, and the initial consensus was that advertising and sales promotion (ASP) expenditures were most likely to account for the high variation in individual

---

* Both the standard deviation and its square, the variance, are tools which summarize the amount of dispersion, or spread, of data.

Table 7.5 Last Year's Sales of Blungits and Advertising
and Sales Promotion Expenditures (ASP) at Ten Stores

Sales (in thousands of units)	ASP (in thousands of dollars)
3.2	0.0
7.9	1.0
1.8	0.0
12.6	2.0
12.1	2.0
0.8	0.0
5.8	1.0
10.6	2.0
16.0	3.0
4.9	1.0

store sales of blungits. Data on the magnitude of the explanatory variable, monthly advertising and sales promotion expenditures, were collected for the 10 stores and are shown in Table 7.5 alongside the number of units sold.

These data are plotted in Figure 7.15 with the regression line that Farr's market researcher fitted to the data. You can look upon the regression line for sales in thousands ($Y = 1.8 + 4.8X$) as defining a *moving arithmetic mean* that depends on the ASP expenditures. Thus, for any level of ASP expenditure, you have some "mean" to use as an estimate. This mean depends on the value of ASP expenditures.* To measure how well the line estimates the observed values, we measure the variance with respect to this moving mean. For each observed value of sales, we can calculate the estimate given by the line. This is done in Table 7.6. The difference between the observed value and the predicted mean is called the *residual.* As you can see in Figure 7.15, the residuals around the regression line of sales are much smaller than were the residuals around the average sales. For any new store, we can decide how much we will spend on advertising and sales promotion and make a better estimate of sales than by just using the mean of all the values.

For example, if we decide to spend $1500 per month on ASP, we predict sales of $1.8 + 4.8(1.5) = 9000$ units per month, rather than 7570 (the mean of all the data). How good is this estimate? Well, you can measure the dispersion of an estimate obtained this way by comparing the variance around the moving mean with the variance around the mean of all the values without taking

---

*And, hence, is technically called the "conditional mean," since it is "conditional" on the value of the explanatory variable.

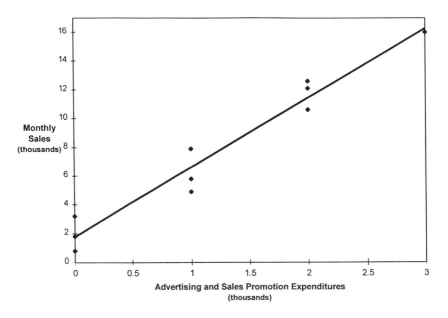

Figure 7.15   Scatterplot of blungit sales and advertising sales promotion expenditures with its regression line.

ASP expenditures into account. The variance around the moving mean is 1200, which you can compare to the variance around the mean of all the data of 26,000.* Thus, we can say that use of the explanatory variable, ASP, has explained $26,000 - 1,200 = 24,800$ of the original variance. Statisticians and researchers usually express this in percentage terms; in this case they would say that "the use of the ASP expenditure as the explanatory variable has explained 24,800 divided by 26,000, or about 95% of the variance."

What does all of this have to do with the correlation coefficient? A lot! The square of the correlation coefficient, R-squared, *is* the proportion of the variance that has been explained by the use of the explanatory variable.† Because of its importance and this usage, R-squared is called *the coefficient of determination,* which is another tool.

Now you have an easy-to-use tool for making your own judgment of the meaning of a particular correlation coefficient when fitting a line to ob-

---

*The computation is not shown here.

† This is an approximation which gets better as the number of data values gets larger. The minor difference between this value and the exact value is not relevant in most practical cases, and we ignore it here to make it easier for the reader to understand the discussion.

Table 7.6 Residuals for Last Year's Sales of Blungits for Advertising and
Sales Promotion Expenditures at 10 Stores as the Explanatory Variable

ASP (in thousands of dollars)	Sales (in thousands of units)	Moving mean	Residual
0.0	3.2	1.8	1.4
1.0	7.9	6.6	1.3
0.0	1.8	1.8	0.0
2.0	12.6	11.4	1.2
2.0	12.1	11.4	0.7
0.0	0.8	1.8	−1.0
1.0	5.8	6.6	−0.8
2.0	10.6	11.4	−0.8
3.0	16.0	16.3	−0.3
1.0	4.9	6.6	−1.7

Note: Calculations from the exact equation $Y = 1.775 + 4.829 X$ have been rounded
to one decimal place.

served data points: Its square is the proportion of variance explained by the
explanatory variable. Thus, for the data of Figure 7.10, the GMAT test score
explains only R-squared = .21 times .21, or .044—less than 5% of the variance in the GPA!* Ninety-five percent of the variance is *not* explained by the
GMAT. The median proportion explained for 20 schools (14) is R-squared =
.35 times .35, or about 12%. In this case, 88% of the variance is *not* explained
by the GMAT.

Where Do You Get Those Ideas?

Two researchers carried out a study of attitudes toward abortion of over 1000
Catholic Mexican-American women living in Los Angeles (13). They divided
the women into two groups, depending on whether their upbringing was in
Mexico or the United States, and performed a complex regression analysis.
Whereas Dr. Nowall and our admissions committees used equations with only
one "explanatory variable"—the divorce rate, SAT score, or GMAT score—
the researchers used several, such as religiosity, husband's occupational prestige, ethnicity, husband's education, own education, age, family size. The last
four explanatory variables can be accurately defined. The others were obtained

---

*In this same school, another study showed a correlation coefficient as high as .4. Thus for this
school, the GMAT explains, at best, about 16% of the variance in GPA. The correlation changes
with time, as does the composition of the student body, faculty, and course offerings.

from questionnaires designed to measure these ill-defined concepts. Despite the inclusion of these other variables, the general idea is the same here as in our earlier examples: to use the data to establish the extent of association of some explanatory variable or variables with a *dependent* variable (e.g., level of education as associated with attitude toward abortion) and to get an equation for that relationship.

To measure how well explanatory variables predict the dependent variable when there are several explanatory variables, we have yet another tool, the *coefficient of multiple determination,* which can be interpreted in the same way as the simple coefficient of determination we have already discussed.

In this case, attitude toward abortion, we have a number of explanatory variables. The values of R-squared for the two groups of women and their several explanatory variables were .15 and .13. To judge these values, Dr. Nowall got R-squared = .83 * .83 = .69 for his analysis of personal bankruptcy and credit card balances, thus explaining about 70% of the variance. However, in the abortion case, only 13% to 15% of the variance was explained by using several variables. Any one variable by itself does not explain much variance. Have these researchers really found a basis for saying what factors enter into the attitudes of these women toward abortion when they can explain only some 13% to 15% of the differences in these attitudes, or is this another poor use of statistics?

EXPERIMENTAL DESIGN

Can a promising growth hormone reverse some of the effects of aging in over-60 males? Does a newly formulated paint last longer than the old formulation? Questions like this are best answered by running an *experiment* in which the investigator exercises control over the subject of interest and the conditions pertaining to the measurement. Good experimental design includes the use of controls, blinds, randomization, systematic allocation to test situations, and many other methods we don't have space to include. These are relatively new tools; the philosophical underpinnings were developed in the 18th century, but experimental design as we now know it is largely a 20th-century development, and still evolving. Much of the recent progress in medicine (particularly new drug and epidemiological successes) is based on the application of these principles. In any given experiment you use several statistical tools.

Color Me Blue

A paint company develops a new formula for exterior house paint which it believes will last longer than other paints. The paint is marketed for two years.

Then the company takes a random sample of purchasers and asks them if they think the paint has lasted longer than other paints they have used. If the results are in favor of the new paint, what do we know? Only the consumers' *perceptions* of the performance of the paint. Important as this is for marketing purposes—and it can be a most important contribution to future profitability—we really know very little about the performance of the paint. Why?

This survey lacks the element of *control*. We don't know if the consumers used this paint differently from the others against which they are comparing it. We don't know if their comparisons are with respect to paints used in different locations (the comparison paint used in the Canadian Arctic, the new version in southern California). We don't know the types of surfaces on which the paint is used or the amount and types of additives, the methods of application, the number of coats, the differences in weather conditions, and so forth. You should be able to add many more factors to this list which could influence the performance of the paint and are not controlled.

What you really want for evaluation is an experiment in which control is exercised over the relevant factors. Most people regard it as just a matter of common sense that you would: Systematically apply the paint in the recommended way to many different but likely surfaces; expose the paint in different localities and different directions in those localities; and measure the performance with some objective measurement such as thickness, adherence, flaking, and so forth. They also understand that there should be *controls,* using earlier or competitive versions of paints treated in the same way. You may have seen such controlled experiments for paint and other surface treatments where patches of the competing surface coverings are made on the same panel of a surface material so that all conditions of exposure and surface are identical.

And even where the measurement is by machine, most people would again regard it as a matter of common sense that the person doing the measurement would not know which paint was which, so as to avoid unintentional bias. When an experiment is run in this manner, we say that it is *blind*, and as we discuss below, there are several kinds of blindness to consider.

Continuing the paint example, you might also be concerned about the location of the comparison patches on the test surfaces. Should our new paint always be in the upper right-hand corner? There might just be some effect which favored or did not favor the upper right-hand corner (such as wind direction or impact of rain). Thus we think in terms of either *randomizing* the placement of the patches, or systematically assigning patches to different locations.

You Don't Have to Grow Old

In 1990, a paper in *The New England Journal of Medicine* reported on an experiment testing the effect of a growth hormone on reversing some of the effects of aging in 12 over-60 males. When interviewed by reporters, the test subjects "gave the hormone rave reviews, boasting of their energy, strength and rejuvenated complexions" (14). Unfortunately, based on a second study of the growth hormone, over-60 males may just have to continue to accept the effects of age on muscle strength, endurance, and mental ability. The preliminary result reported in 1990 was used to guide a subsequent experiment employing the proper array of experimental tools. In the second experiment, reported in *The Annals of Internal Medicine* in 1997, there were 52 test subjects who were split into two groups to provide a control group, meeting our first requirement for this type of experiment. In addition, the study was *double-blinded*; neither the researchers nor the patients knew who was getting the treatment and who was getting a placebo. The results were not as promising.

As the director of the second study said, "The [initial] study wasn't blinded... [the patients] knew they were taking growth hormone, and may have believed in it so strongly that the power of suggestion made them feel better. . . ." A collaborator on the 1990 study also said, "I think those men were expecting to feel better, and they did. They became euphoric." She also said that some of the patients may have been led by reporters' questions to exaggerate a bit.

This is something to bear in mind the next time you hear about the beneficial effects of a food, drug, appliance, or physical regimen that is supported only by anecdotal accounts. Magical thinking does not substitute for experimental proof.

SUMMARY

Is Nothing Good Enough for You?

Does this recitation of flaws related to improper methodology in statistical experiments give you the feeling that no one will ever be able to perform an experiment that meets our demanding standards? Of course not! We don't expect or demand absolute perfection, agreeing with the one of the great statisticians of all time, Jerzy Neyman:

> The tests themselves give no final verdict, but as tools to help the worker who is using them to form his final decision. . . . What is of chief importance in order that a sound judgment may be formed is that the method adopted, its scope and its limitations, should be clearly understood . . . [15].

The great majority of statistical experiments are not misuses and stand up to the critical investigation to which they are exposed by their publication. Many Nobel prizes have been won by investigators who have used their tools well.

We learn about new directions of study from less than perfect as well as from superbly designed experiments. A large part of the remarkable improvement in public and private health is the result of sound statistical experiments which are a part of an ongoing process of searching for answers.

Good work *can* be done. It is a statistical crime to not try.

8
Faulty Interpretation

Easily doth the world deceive itself in things it desireth or fain would have come to pass.

—Montaigne

Facts, or what a man believes to be facts, are delightful. . . . Get your facts first, and then you can distort them as much as you please.

—Mark Twain

INTRODUCTION

Faulty interpretation is a major misuse of statistics. In this chapter we discuss faulty interpretation, how it happens, and the consequences.

We count, rank, and measure to give substance to our observations. This "substance" we call numbers, which can be correctly tabulated or summarized using arithmetic. But the numbers alone are not sufficient for accurate interpretation. We must know to what they refer and what they mean. *Definition* is how we decide what is counted or measured, and faulty definition is one of the factors in misinterpretation. *Conflict* is another factor in faulty interpretations. When different organizations or statisticians reach different conclusions using the same information, the conflict itself can create faulty interpretations. *Lack of knowledge* of the subject matter, as discussed in Chapter 3 (Know the Subject Matter), is another source of incorrect interpretation, and is not covered in this chapter. Another cause of faulty interpretation is *blatant misinterpretation*—data are misinterpreted not through ignorance, but simply because the obvious is ignored. Blatant misinterpretation is related to lack of "thinking" (see Chapter 12). *Definition* and *lack of knowledge* were discussed earlier in another context. In this chapter we apply some of these concepts to the problem of interpretation.

DEFINITION

All statistics—both the data and measures derived from the data—are *artifacts*. The following simple examples illustrate what we mean by this.

How Big Is Your Apartment?

Some statistics are conceived to be integers (whole numbers), such as the number of rooms in an apartment. At first glance, the number of rooms in an apartment appears to be an integer, such as 1, 2, 3, etc. But we have all seen advertisements for 1½-, 2½-, 3½-room apartments, so that our concept of discrete is broadened in this case to include integers with halves.

Occasionally, because of the desire to simplify, or ignorance, an investigator may reduce all numbers of rooms to integers for convenience. Then rules must be made to determine the number to be assigned to each type of room. What is the number assigned to a foyer, a walk-in closet, or a combination living and dining room? How is a "foyer" or a "combination living and dining room" *defined?* The assigned number of rooms then becomes an artifact, a consequence of the definition. Without explaining the definition, the resulting statistics may mislead others.

How Old Is Your Child?

Other statistics are conceived as continuous; that is, taking on real number values. Age is such a variable. No person has an age that is "exactly" a unique value. Age is a measurement on a time continuum, which you can measure as precisely as you wish, depending on the nature of your clock. However, for statistical purposes, we usually measure age in years. To do this, we make arbitrary definitions about where the line is drawn to obtain integer values for age. The most common usage is age at last birthday, although sometimes age at *nearest* birthday is used. Prior to World War II the Japanese automatically aged everyone one year on the first day of the New Year, regardless of when the person was born. "Should a child be born on December 31st, or on any day of the year prior to that date, the infant on January 1st of the year following is regarded as being two years old" (1). For example, a baby born up to Dec. 31, 1930, automatically became two years old on Jan. 1, 1931.

Once again, we see that the definition creates statistics as artifacts. Thus, there are no uniquely correct data!

Problems of Definition

How Poor Are Your Statistics?

Giving aid to the poor is a continuing human issue. But in order to give aid effectively and efficiently, you must first define what is meant by "poverty."

Only then can you count the number of people who live in poverty—the poor. You must also know the change in the number of poor with time, in order to evaluate assistance and employment programs. Political reputations and human suffering rise and fall with the answers to these questions.

In the United States, a family is categorized as poor depending on the family's annual before-tax money income. Those receiving less than this specified amount are "below the poverty threshold" and counted as poor. From time to time, government officials and some members of Congress have proposed including the cash equivalent of noncash benefits (such as food stamps and medical assistance) as part of the family income.

If such a definition were used, a family could rise above the poverty threshold through its own sickness. If all members of a family were ill and receiving medical aid, this amount of noncash assistance, added to their regular cash income, could lift them out of the "poor" status, and this family would be eliminated from the welfare rolls. But this change in their status could also mean that certain types of assistance available to the "poor" would now be unavailable to them. Thus, by practical and social standards, they could become even poorer as a result of illness! A real Catch-22, as shown in Figure 8.1.

Figure 8.1 A newspaper cartoonist agrees with us on the consequences of including benefits to the poor in their income. Wasserman, for the *Boston Globe*, copyright 1985 by Los Angeles Times Syndicate. Used by permission.

If the government adopted this change in the definition of poverty, the number of "poor" would drop and politicians could argue that poverty had been reduced. Even without this self-serving misuse of the revised definition, such a change would lead to underestimating the need for aid when policy decisions were being made.

As we discussed in Chapter 4 (*In the Number of Poor People?*), the Joint Economic Committee of Congress recognized the importance of this statistical artifact and the need for a rational process to determine the threshold of poverty. Their interest led to a comprehensive study by the Panel on Poverty and Family Assistance under the direction of the National Committee on Statistics. The committee recommended a new measure subject to three criteria:

> First, a poverty measure should be acceptable and understandable to the public. Second, a poverty measure should be statistically defensible. . . . Third, a poverty measure should be feasible to implement with data that are available or can fairly readily be obtained [2].

These conditions make it clear that the measure of poverty in the U.S. is an artifact, to be determined by political and scientific processes.

Would a Change in Definition Make Any Difference?

The answer to this question depends on the subject matter and the degree of precision you seek. We can illustrate this with the measure of unemployment used in the United States. First, let's briefly review the origin and nature of today's employment and unemployment statistics and see how an artifact is created.*

During the Great Depression of the 1930s, the U.S. government enacted a number of provisions to assist the destitute, such as the Works Progress Administration (WPA), the Civilian Conservation Corps (CCC), and so forth. The clamor was for jobs, and the government was asked to create more jobs. Lawmakers and the executive branch alike were faced with the question: How many people are unemployed and need jobs? Only with this information could the government decide how much money and effort to allocate to job creation programs. No one had any answers, for no one knew with any reasonable degree of certainty the number of unemployed.

Consequently, the WPA embarked on a research program to define "employment" and "unemployment" and to learn how to count the number of

---

*For further information see Refs. 3 and 4 and any current issue of the monthly publication of the U.S. Bureau of Labor Statistics, *Employment and Earnings*.

people in each category. The result was a definition of what is now called the "labor force." The first major classification included only people involved in the money economy. It was felt that there was no point counting housewives or volunteer workers (such as people who assist a religious institution or work in a museum without pay, and so forth) since Congress was not considering appropriating aid for such people. The government's plan was to use separate programs to assist those who were unable to work, were too young or too old, or were not participating in the money economy.

How do you count these people? Without going into details, the reasoning was generally as follows:

1. The U.S. labor market is based on competition for jobs.

2. A person enters the labor market when that person competes for a paid job.

3. People who are employed have successfully competed in the labor market and thus are one part of the labor force.

4. Those who do not have jobs but are seeking them are competing for jobs in the labor market and form another part of the labor force.

5. People who do not compete for jobs, or are in institutions such as jails or hospitals and could not compete for jobs, are excluded from the count of the labor force.

When these concepts were first established, members of the armed services were also excluded from the labor force on the grounds that there were too few of them to matter, and as they were already "employed" by the U.S. government they could not compete for jobs to be underwritten by Congress.

Thus, the labor market is defined as those people who are employed plus those among the civilian, noninstitutional population who are actively seeking jobs. This sounds like a simple definition, but is it simple to put into practice? How can this information be obtained? By conducting a survey and asking people. How often do you conduct the survey? Congress and the Administration wanted information monthly. Therefore, the survey was conducted monthly. What should the time reference be? If you ask someone if he or she is employed or seeking work, you must specify some precise period. The period selected was one week in a month, in particular that week which includes the 19th day. If the specified period is longer or shorter than one week, the numbers of employed and unemployed vary, but we do not go into that here. In any event, one part of the artifact being constructed is the length of the reference period.

Before the labor force artifact can be completed, a number of additional decisions must be made. Should the survey ask everyone or only those of some specified age? During the exploratory work of the 1930s, researchers realized that children under age 14 were unlikely to be either employed or unemployed. Thus the survey was initially limited to the population aged 14 and over. Subsequently it was found that very few 14- and 15-year-olds were reported as either employed or unemployed, so the lower boundary of age was raised to 16 years.

The survey designer had many more decisions to make before the artifact could be finished. For example, what about people who work on a family enterprise, such as a family farm or small business, for which they receive no money? Should they be counted as employed or not? The decision was that if they worked 15 hours or more per week, they would be defined (and thus counted) as employed. Today, we might also include welfare recipients on part-time "workfare" assignments.

Another question concerned the reference week. During this week some people were both employed and unemployed. In which column did they belong? Since any one person could be counted as only one or the other, the survey designers decided that such people should be counted as employed.

Determining the official "unemployment rate" involved innumerable decisions. Some of these decisions, and the resulting procedures, have been modified considerably over the years, but the basic framework of the labor force artifact remains: *The labor force is those who are employed plus those who are seeking work.* The rest are not in the labor force.

The U.S. Bureau of the Census conducts this monthly survey on a sample of the households in the United States, and the U.S. Bureau of Labor Statistics analyzes the data and publishes the reports.*

In light of the decisions which led to the present labor force artifact, we can ask: What would happen to the unemployment rate if a person who was both employed and unemployed in the reference week was classified as "unemployed" instead of "employed," as is now done? In 1984, the official unemployment rate for the population aged 16 and over was 7.5%. There were 105 million employed; included in this figure were 5.7 million who were employed but worked only part time for economic reasons. It can be argued that they should be classified as unemployed during the time they were involuntarily not working. In addition, there were 8.5 million who were officially unemployed. If the 5.7 million who worked part time only because full-time work was not available were added to the 8.5 million, then we would have 14.2 million unemployed, or 12.5% of the civilian work force (5, Tables 1 and 7). Which unemployment rate do you prefer, 7.5% or 12.5%? Or something in between?

Another change has occurred: the "labor force" has been redefined. For over 30 years, until 1983, the unemployment rate has been shown as a per-

---

*An example of such a report is Ref. 4.

centage of the total *civilian* labor force. But if you add the number of *military* personnel (who are, after all, employed and working) to the civilian labor force to get a total labor force, the statistics change.

When the labor force measurement procedures were designed, the number of military personnel was proportionately smaller than it is today, and it was thought that only the civilian component of the labor force was relevant. In 1983, military personnel were added to the labor force. The result is what you would expect: the unemployment rate was immediately reduced!

For women and older men, this change made no difference, since so few are in the military. For younger men, the decrease in unemployment rate is significant. Including the military in the labor force "answers" the cries of the media and politicians that the unemployment rate among young men aged 18 to 24 is too high: When military personnel are included, the young-male unemployment rate drops. For example, the unemployment rate in 1982 for nonwhites between 18 and 24 years of age drops from 32% to 28% when military personnel are included. (The rate for white youth of the same age group drops only 1%, from 16% to 15%, because relatively fewer of them serve in the military forces [6, Table 3].)

Even these (reduced) rates were too high; nothing was accomplished by changing the definition of the labor force except a change in the interpretation. Unfortunately, however, this "paper" reduction in unemployment rates could be touted by politicians as an improvement and result in a drop in federal allocations for programs for the unemployed.

Note: In 1994 another change occurred in the definition of the labor force: military personnel were no longer counted as part of the civilian labor force. In 1996, unemployment came to 5.4% of the labor force. And consider what effect removing military personnel can have. In 1995, unemployment was 5.6%. If the employed force included military personnel, the number would have dropped quite dramatically, to 4%. If there was a political reason for the 1984 change which resulted in lower unemployment figures for that period, was there another political reason for the reversal in 1994?

CONFLICTING NUMBERS

The Case of India and Its Food Supply: Who Knows?

Whether India was able to grow sufficient food for its own population was an issue of international importance in the 1980s. According to M. T. Kaufman in an article in *The New York Times,* the answer was "yes" (7). The article begins: "It has been four years since the last shipload of foreign wheat arrived in India."

He then continued with information based on a World Bank analysis and Indian government statistics: "World Bank analysts point out that as grain production has increased, and prices have dropped, consumption has grown to approach minimal levels of nutrition" (7). The increased production of grain (or all foods, the article is not completely clear here) reportedly came about from the introduction of new agricultural methods, the maintenance of stable farm prices, and other governmental measures.

We ask: How minimal is minimal? According to the World Bank, in 1980, the daily calorie supply per person in India was 1880 calories, 87% of the daily requirement. The daily calorie supply per person of 1880 calories is an average, which means that many people consumed more than 1880 calories and many consumed less. We suspect (but cannot verify) that the average is 1880 calories because the fortunate few who eat a great deal more than 1880 calories offset the larger number who consume less than 1880 calories. Thus, there could be a large number of people who do not approach 1880 calories per day, which is still only 87% of the daily requirement. If you compare the U.S. average calorie consumption of 3860 with the Indian figure of 1880, you see a shocking difference—the average Indian caloric intake was less than half of the U.S. average at the time (8, Table 24).

What was the average consumption of calories per person in India before 1980? Information for 1960 leads us to believe that there had either been no change over the preceding 20 years or a small decrease. In 1960, the calories consumed in India per person per day was reported to be 1990 (9, Table 1240), which is only 6% higher than was reported for the year 1980. Because of the likely error inherent in the average calorie consumption data, we are inclined to take "no change" as the verdict.

Let's look further into the available information on food production in India. The United Nations Food and Agricultural Organization (FAO) gives data (10) which conflict with the *New York Times* story (7). The indexes of food production per person in Table 8.1 show year-to-year fluctuations but no steady improvement in the food supply.

Who is right, the World Bank or the FAO? To try to answer this question, we look more closely at the statistics. If the World Bank's numbers refer to total food production, then they are obviously different from the per-capita figures cited by the FAO. If the World Bank numbers refer only to grain, then grain production could have increased while total food production per person did not. This would have occurred if production of nongrain foods was cut back at the same time, which is not an unlikely event, since resources such as land are often diverted from one agricultural crop to another. Another problem is that FAO sometimes publishes slightly different values for index numbers for the same reported year. For example, the per-capita food production index for 1979 was reported as 97 in the 1981 volume and as 101 in the 1983

Table 8.1 Indexes of Indian Food
Production (per person)

Year	Index
1969–71	100
1976	98
1977	107
1978	108
1979	101
1980	102
1981	108
1982	102
1983	114

volume (10). We assume that the latest values (the 1983 figures) for 1979 are likely to be more nearly correct and have used the most recent values in our discussion.

Further, statistics on agricultural production, regardless of the agency that publishes them, often contain errors of unknown size. We give an example in the next section.

Based on our examination of the reported statistics, can we say any more than that India was just holding even? If India's ability to feed itself was an important issue, then didn't these two prestigious organizations owe us a reconciliation of their discrepancies? And shouldn't the author of the *Times* article have pointed out that alternative and apparently contradictory information exists? Shouldn't the report also have defined whether caloric consumption included only grain or *all* foods? In the end, all the reader can do is to *stay on guard* and not automatically believe all published interpretations.

Agricultural Production in El Salvador

The effect of land reform in El Salvador was an important political issue in 1983. Illustrating an excellent example of investigative reporting, R. Bonner of *The New York Times* reported on agricultural production and presented all the information available (11). Three concerned groups—the Salvadoran government, a U.S. government agency (the Agency for International Development), and the University of Central America (Center for Documentation and Information)—made statistical estimates of production and its change over time. The statistics on 1980-81 coffee production, shown in Table 8.2, are a good example of how production estimates can differ.

The differences speak for themselves. But what about how these estimates are reported? Not only do the three agencies differ considerably, but

Table 8.2 Conflicting Estimates of Coffee Production in El Salvador

Source of estimate	Production (tons) 1980–81 coffee
Salvadoran government	25,950
U.S. government	20,204
University of Central America	16,951

how can we give credence to estimates that attempt to give an impression of exactitude by reporting numbers such as 20,204 or 16,951? In our view, the only reasonable reports are 20,000 and 17,000.

BLATANT MISINTERPRETATION

At the start of this chapter, we defined blatant misinterpretation—what the dictionary calls "glaringly evident" (12, p. 97)—as the result of the reporter or researcher ignoring the obvious. But not everyone sees the obvious until it is pointed out, when it becomes glaringly evident. This is why all of us have made such misinterpretations at one time or another, and it is through the open exchange of information (as in Letters to the Editor and scholarly commentary) that we slowly reduce the number of blatant misinterpretations.

Making Sense Out of Numbers

The Quality of Teeth

In an article in the *New York Statistician,* Mark L. Trencher brought to our attention a statement by *U.S. News & World Report* on tooth decay:

> The rate of tooth decay among school children has dropped roughly one-third in 10 years, indicates a recent survey by the National Institute of Dental Research. This nationwide sampling . . . showed that 37 percent . . . had no decay. A comparable sampling 10 years ago showed that only 28 out of 100 children were free of decay—a difference of 32 percent from the current population [13].

The "32 percent" is the reduction in percentage points of the number *without* cavities obtained by dividing the reduction in percentage points (37% minus 28%) by 28%. But the article is about the reduction of those *with* decay!

The reduction in the rate of those having tooth decay is only 12.5%. To get this value, first you find that the percentage having tooth decay 10 years ago was 72% (100% minus 28%) and that it is now 63% (100% minus 37%). The reduction in the rate for those having tooth decay is 9 percentage points (72% minus 63%). Divide this reduction in percentage points of those having decay by the rate for 10 years ago and you get 12.5% (9% divided by 72%) as the percentage reduction of those having tooth decay. A reduction of 12.5% in the rate of tooth decay among schoolchildren is a cause for joy and encouragement, but it is *not* a reduction of "roughly one-third."

Percentages seem to invite misuse, perhaps because they require such careful thinking. We discuss problems with percentages in more detail in Chapter 12 ("Thinking").

There Were Plenty of Jobs for the Unemployed—Or Were There?

At a news conference in 1982, then President Ronald Reagan suggested that unemployed people should look at the help-wanted sections of their newspapers, where they could find many opportunities. President Reagan said (in the Jan. 17, 1982, issue of the *Washington Post*): ". . . there were 24 full pages of classified ads of employers looking for employees."

However, according to the AFL-CIO's *Revenews:*

> Actually there were 26 pages of help-wanted ads.*. . . But of these 26 pages, four were devoted entirely to the engineering profession, three to nursing, three to computer programming, two to secretarial work, and two to accounting and bookkeeping. . . . [A total of] no more than 3,500 job openings, but there are now more than 85,000 unemployed people in the Washington Metropolitan Area [14].

Twenty-four (or 26) pages of classified ads for help may seem ample. But the total number of opportunities was small compared to the number of the unemployed, and most of the opportunities were for skills which many of the unemployed did not have!

Choose Your Drink

Sometimes you need statistical analysis (but not necessarily a great deal) to verify whether or not an interpretation is correct. In 1981, the Joseph Schlitz Brewery Company staged live taste tests of its beer against competing brands

---

*This is one of those rare cases where an error in basic data was not in a direction supporting the speaker's contention (in this case, of a plenitude of opportunities). This supports the view we express in Chapter 1 that you should be cautious in attributing misuses or bad data to deliberate actions. The fact that most such errors tend to support the speaker's or author's views does not mean that they deliberately committed these errors. We all make errors.

during the Super Bowl and two National Football League playoff game telecasts. In each of these three surveys, 100 self-identified regular users of competing brands of beer (Miller, Budweiser, Michelob) were given blind, paired comparison preference tests with respect to Schlitz's beer.

In a "paired comparison preference test," the test subject compares two products and states a preference. A test is "blind" when the test subjects do not know which of two products they are testing when they make their choice. It is well known among market researchers that a fundamental problem with such tests is that, occasionally, some test subjects have expressed a preference for one product over another when the two products were identical (15)!

If the two products are definitely different, such as steak and carrots, a preference may be taken as real. But if a paired comparison of two *similar* products results in an even split of preferences, we have no way of knowing whether this is because of true preferences or because of the inability of the test subjects to taste a difference between the two products. They may simply be guessing.

The numerical results are shown in Table 8.3. In reporting the results, the announcer emphasized the high percentage of testers (ranging from 37% to 50%) who preferred Schlitz to their usual brand. Unfortunately for the Schlitz Brewery Company, the message may not be that beer drinkers prefer Schlitz's beer to their regular brand, but rather that beer drinkers cannot tell one brand from another when they make a blind paired comparison. This is an important message for those concerned with the marketing of beer, but not the same as the message that the brewery wanted its television viewers to hear when it announced the results.

To evaluate these results, we hypothesize that the results come from the inability of the test subjects to tell the beers apart, and we judge the results by their deviation from this hypothesis.

If beer drinkers can't discriminate between beers by taste, then we expect that blind paired comparisons will show 50% of the test subjects (or 50 out of 100) stating a preference for Schlitz.

Using this hypothesis, a statistician can compute the probability of getting the observed results. If the drinkers were unable to tell the difference between the beers, we expect results of 50% and 50%, an equal split showing

Table 8.3 Results of the Schlitz Taste Test

Competing Brand	Preferred Schlitz (%) to:
Miller	37
Budweiser	48
Michelob	50

no preference. The results obtained in this case (48% and 50%, respectively, expressing a slight preference for Schlitz) are consistent with a conclusion of no preference. This is an occurrence that is extremely likely to happen if the drinkers have no preference between these two beer brands. To evaluate the result for the comparison with Miller, you can easily compute the probability of occurrence to be less than 1%; the drinkers do show a preference.

The statistics give no reason to believe that the Michelob and Budweiser results are other than guesses, whereas there is strong evidence that the majority of Miller drinkers prefer Miller. This is a different interpretation from what Schlitz reported. For the television viewer, the reasonable conclusions based on this test are:

1. Those who are regular drinkers of Budweiser and Michelob can't taste a difference from Schlitz.
2. Those who are regular drinkers of Miller can taste a difference between it and Schlitz.
3. Most regular drinkers of Miller prefer Miller, but about one-third prefer Schlitz by taste.

As far as we can tell, this test was well designed and executed, but we regard its interpretation as a misuse. In a controlled situation, a market researcher could make excellent use of this test by digging deeper: Who were the individual Miller drinkers who preferred Schlitz and how did they differ from the other Miller drinkers? Who were the Michelob and Budweiser drinkers who chose Schlitz in the comparison? Could they tell a difference, or were they in fact unable to discriminate? If they preferred the taste of one beer over another, why? If they were unable to discriminate, why? Those market researchers who use (as most do) good basic data and sound statistical methodology, paired with correct and careful interpretation, can make better product decisions.

One Giant Logical Leap for Mankind: Unfortunately, a Misstep

Dr. Stephen Kellert (of Yale University) and Dr. Alan Felthaus (of C. F. Menninger Memorial Hospital) examined the association between childhood cruelty toward animals and aggressive behavior in adult criminals and non-criminals, noting: "The research literature appears to suggest, thus, that childhood cruelty toward animals may operate as part of a behavioral spectrum which is associated with violence and criminality in adolescence and adulthood" (16, pp. 5–6).

Three groups of males were examined: aggressive criminals, nonaggressive criminals, and noncriminals. The two criminal populations were drawn from federal penitentiaries, and noncriminals were chosen at random from areas near the penitentiaries. Childhood behavior of these individuals toward animals was determined through interviews with the test subjects, parents, and siblings.

There are many methodological questions to be raised in the conduct of this study, which the authors themselves discussed and either dealt with or took into account in their analysis. Their methodology is not our concern at this time, and we will assume the validity of their conclusion:

> This paper has reviewed a number of results from a study of childhood cruelty toward animals, motivation for animal cruelty, and family violence. The strength of these findings suggests that aggression among adult criminals may be strongly correlated with a history of family abuse and childhood cruelty toward animals. . . . This [sic] data should alert researchers, clinicians, and societal leaders to the importance of childhood animal cruelty as a potential indicator of disturbed family relationships and future antisocietal and aggressive behavior [16, p. 33].

In short, a child who displays cruelty toward animals and becomes a criminal is more likely to be an aggressive criminal or antisocial than a child who is not cruel to animals. What action might be taken as a consequence? As the authors suggest, the behavior of children toward animals might be used as an indication of future problems. That much, and no more, is a legitimate claim, as derived from their investigation and analysis.

But the authors now make a giant—and invalid—logical leap: "The evolution of a more gentle and benign relationship in human society might, thus, be enhanced by our promotion of a more positive and nurturing ethic between children and animals" (16, p. 33). The authors have shown that children who were cruel to animals are more likely to be aggressive adult criminals, but they have *not* shown that the promotion of kindness to animals in a child will result in that child being less likely to become an adult criminal. Association is not cause (as we discuss in Chapter 7), although it may well be that there is an underlying causal factor—the authors themselves mention several—that leads to both childhood cruelty to animals and adult aggressive criminality.

To be able to say that the "promotion of a more positive and nurturing ethic between children and animals" will reduce aggression in adult criminals, you must run a controlled experiment (as discussed in Chapter 7) in which two or more groups of similar children are exposed to "promotion" and "nonpromotion" of kindness to animals and then followed through their adult lives. (Most researchers use identical twins in studies such as this.)

For now, the jury is still out, despite these researchers' good work and intentions.

STATISTICAL SIGNIFICANCE: MISUSE BY MISINTERPRETATION

Having failed to establish himself as a financial consultant based on his analysis of personal bankruptcies (discussed in Chapter 7), Dr. Nowall looked around for another field of endeavor. Sensing that there was even more interest in health care in the United States than in personal bankruptcies, he decided to look at the performance of hospitals. In the managed care environment, an important indicator of service to the patient is the mean length of time spent waiting to be called for admission processing: the time from when patients walk in the door to when they are called to give their information to the admitting clerk. In the competitive health care environment, it is to the hospitals' advantage to reduce the mean waiting time.

Dr. Nowall decided to try for a headline on the performance of the two major hospitals, A and B, in his city. From his limited experience and reports of patients, he believed that the mean waiting times for the two hospitals were different. He reasoned that if, using some sophisticated methodology, he could show this, it would make a good story and help to qualify him as a health care analyst.

It seemed simple enough to find out if there was a difference in the mean waiting time between the two hospitals. If he could sit in the waiting room of each hospital for 24 hours a day, for a full week, he could eliminate any possible effects of different waiting times on different days. Of course, if he had the time and patience to find *all* the waiting times for each hospital (the population of waiting times), he could just compute the population means and know for sure which took longer. But clearly this was impossible. He was going to have to estimate the difference between the mean waiting times. No problem, as he recalled from his statistical training; all he had to do was to set up a *hypothesis test*. With a hypothesis test, he could take a random sample of waiting times at each of the hospitals and see if the samples supported the hypothesis that there was no difference.

The rationale of hypothesis testing is well known and is used in many fields such as science, engineering, and drug testing. To use a hypothesis test to determine if there is a difference in the waiting times, Dr. Nowall set up a *null hypothesis* that states that there is no difference in the population mean waiting times of the two hospitals, A and B. He gets a random sample of waiting times from each hospital, computes the mean of each of these samples, and compares them. This gets tricky because these are the results of randomly chosen samples, and if the sample mean for hospital A is different from that for hospital B, it might just be the "luck of the draw." The apparent confirmation of Nowall's belief that hospital A has a higher or lower mean waiting time than hospital B might just be fortuitous.

The way statisticians have been dealing with this issue for over 150 years is to measure the probability that the result obtained could have occurred by chance if the null hypothesis were true (there is no difference in the population means of the two hospitals). This probability is denoted P or p and called the "level of significance." It is quite common to say that a result is not *statistically significant* if the probability of its occurrence by chance is less than 5%. Of course, when you reject the null hypothesis that there is no difference because your P value is 1%, you are taking a 1% chance of being wrong.*

Dr. Nowall wants to do a good job, so he gets a sample of 300 waiting times from each hospital. When he analyzes the results, his estimate of the difference in waiting times between the two hospitals is one minute, and it is in the direction he suspected: the sample mean of the waiting times at hospital A is 1/4 minute larger than that of the mean of waiting times at hospital B. Since the P value for this result (a calculation that he performed using his computer) is 3%, he was able to announce that there was a "significant difference" in mean waiting times in the two hospitals. This was enough to attract considerable attention to his finding, and hospital A was negatively affected in the health care marketplace of the city because people hearing this statement interpreted "statistically significant" as meaning that the difference was important.

Yes, this was a *statistically* significant result, in the sense that the observed difference was relatively unlikely (4%) to have occurred by chance. But the key question is whether the observed difference reflected a *practical* significant difference. If not, it certainly is not an important finding. The sample mean times were 16 and 16-1/4 minutes, respectively. Is this a *practically* significant difference? Is it an important difference?

It is in the nature of the rationale involved that the larger the sample size, the greater the statistical significance (lower and lower P values). The larger the sample, the less the likelihood of *any* difference occurring by chance. The following comment on this subject appears in a letter from W. Edwards Deming, one of the great statisticians of the 20th century:

---

*If this short and elementary summary of the rationale of hypothesis doesn't seem convincing, consider this analogy: Your dog has eaten your neighbor's nasturtiums. You want to investigate your neighbor's feelings toward you after this incident, and approach the neighbor while she is watering her garden. Your null hypothesis is "neighbor bears no resentment toward me." The neighbor turns, faces you, and sprays you with water from the garden hose. How likely would this occurrence be if the null hypothesis were true? Rather unlikely, so you reject the null hypothesis that the neighbor bears no resentment. On the other hand, you can't be certain. The neighbor might have been startled and turned the hose toward you by chance. The probabilities determine your conclusions.

Someone came in from the Animal Husbandry Department [of the University of Wyoming] to announce . . . that the fibres on the left side of a sheep, and those on the right side, were of different diameter. Dr. Bryant asked him how many fibres they had in the sample and they told him 50,000. This was a number big enough to establish the fact that there was a difference between the left side and the right side, a fact that ANYBODY COULD HAVE ANNOUNCED BEFORE THE EXPERIMENT [17].

It is a pity how many foolish calculations are made on significant differences. As Tom Houston said in *BYTE*, "The statistical meaning of the word 'significant' is distinct from the popular sense of important or meaningful. . . . Unfortunately the colloquial or ambiguous use of 'significant' sometimes obscures *carelessly written* reports. . . . [our emphasis]" (18).

The foolishness goes even further. Some people, including those in the legal profession or serving on juries, interpret increasingly higher levels of statistical significance (lower P values) as measures of increasing importance. Protecting the public against this misuse can have great consequence, but sometimes the statistical profession itself feeds the Misuse Monster. A best-selling statistics textbook for business suggests the following phrases to use to "describe your results" (19, p. 354):

Not significant	Not significant at the conventional 5% level
Significant	Significant at the conventional 5% level
Highly significant	Significant at the 1% level
Very highly significant	Significant at the .1% level

It would be much better to stick to reporting the P value (level of significance) and the sample size and let the readers draw their own conclusions about the *practical* significance of a result which is *statistically* significant. As textbook authors Moore and McCabe say:

Other misuses involving interpretation abound. For example:
The 5% level ... is particularly common. Significance at that level is still a widely accepted criterion for meaningful evidence in research work. *There is no sharp border between "significant" and "insignificant," only increasingly strong evidence as the P-value decreases.* There is no practical distinction between the P-values 0.049 and 0.051. It makes no sense to treat [a significance level of 5%] as a universal rule for what is significant [20, p. 486].

SUMMARY

Misinterpretation arises from many causes, some of which are discussed in this chapter.

Some Guidelines for Readers (and Writers)

What are the definitions? Definitions by which data are collected or tabulated may not lead to answers to the questions being asked. Always try to find out the definition used to categorize, which good writers and authors will give in the text of their articles and reports.

Changing times. Definitions may change with time. Determine if the definition has changed.

Know the subject. Knowledge of the subject is relevant to collecting data and drawing conclusions. To avoid being misled by a misuse of statistics, know the subject or consult an expert.

Baseline values. Incorrect baseline values lead to misleading results in comparisons, as for example in the discussion concerning tooth decay in this chapter. Find out what the baseline is and if it is consistent. You can usually determine the base by a simple computation, as we have done in this chapter and elsewhere in this book.

Spurious precision. Since spurious precision gives false confidence in interpretations, round published statistics to numbers that make sense in view of the way in which the data were collected and processed.

Differentiate between practical and statistical significance. Once again, we quote from Moore and McCabe, *Introduction to the Practice of Statistics:*

> The remedy for paying too much importance to statistical significance is to pay attention to the actual experimental results as well as to the *P*-value. Plot your data and examine them carefully. Are there outliers or other deviations from a consistent pattern? A few outlying observations can produce highly significant results if the data are run blindly through common tests of significance. Outliers can also destroy the significance of otherwise convincing data. The foolish user of statistics who feeds the data to a computer without exploratory analysis will often be embarrassed. Is the effect you are seeking visible in your plots? If not, ask yourself if the effect is large enough to be practically important [20, p. 487].

9
Surveys and Polls: Getting the Data

I am not bound to please you with my answer.

—William Shakespeare

It is better to know nothing than to know what ain't so.

—Josh Billings

INTRODUCTION

You survey to know the general state of affairs: opinions, events, attitudes, intentions, ownership, habits, purchases, demographic characteristics, and so forth. *Observation* provides the basic data for a survey. You count cars passing an intersection, report about the condition of bridges, and ask questions of consumers, voters, and inhabitants. Surveys have a long history and their types and purposes are diverse.

There are many different kinds of surveys, but today, those in which we pose questions to people and then record, analyze, and interpret their responses have the most impact in the political arena. Because political surveys are so important, it is the only type we discuss, and in this chapter we give examples of misuses of statistics in such surveys. We cannot give examples of all possible misuses, for they would fill several volumes. However, we do illustrate the kinds of reasoning you can use to examine such surveys for these misuses.

A survey has three basic parts: the questions, the responses, and the analysis. The final product—the report—often appears to be scientifically correct whether it is or not. Professional pollsters and surveyors are aware of the pitfalls inherent in this process and usually try to avoid them, but even the professionals don't always succeed. Nonprofessional pollsters and surveyors are often unaware of the pitfalls and can make so many errors in polling or surveying that some of their work qualifies as a megamisuse.

FIRST PRINCIPLES

To Census or to Sample?

A survey may include all of the population of interest or only a part of it. Statisticians call the entire population of interest the *population* or the *universe*, and a part of the population a *sample*. If the survey includes all of the population, it is called a *census*. If it includes only a part of the population, it is called a *sample survey*. Some surveys include all of the population for one set of purposes, together with a sample of the same population for other purposes, as in the case of the 1990 U.S. Census of Population and Housing which we discuss later.

Whether we make a census or a sample survey, we pick a group of individuals and ask them questions. Those who respond become *respondents;* those who do not respond become *nonrespondents*.

The U.S. Constitution requires a count of the population every 10 years to allocate members by states to the House of Representatives. Legislation has added the "one person, one vote" rule, which requires that all congressional districts within a state be as equal in population as is possible. The U.S. Census of Population must first satisfy the state by state counting requirement of the U.S. Constitution. Then, to satisfy the "one person, one vote" requirement, the Census must get a complete count of everyone in every county, city, and smaller defined areas. For these purposes, a complete, nationwide count is conducted every 10 years. The 1990 U.S. Census asked only seven population questions and seven housing questions of every household and its members.

There is no Constitutional requirement to collect information on other subjects, and where other information is sought, a sample is generally used. In the 1990 U.S. Census of the Population, the additional data collected were responses to about 57 questions (which included the 14 housing and population questions asked of every household). "About" 57 questions? Yes, the number of questions is approximate because some questions had several parts and the number depends on what parts are counted as separate questions.

The *sample* questions were asked of every sixth housing unit (about 17%) in most of the country. If an administrative unit—county, incorporated region, or other local division—contained fewer than 2500 people, the sample questions were asked of every other housing unit (about 50%) (1, p. 1). The answers from these samples are projected to the population of each sampled area by using the ratio of the sample to the counted population of the area, a process which is subject to sampling error (see Chapter 10).

In 1980, the census questionnaire also included a supplementary sample of the Native American population. All Native Americans (American Indians, Eskimos, Aleuts) living on federal or state reservations, or the historic areas of Oklahoma, were asked additional questions. Native Americans living elsewhere were not included in this sample. The Census Bureau will not include this supplementary sample in the year 2000 census.

How Big Is Big?

The size of surveys varies greatly, from a dozen respondents to hundreds of millions, as in national censuses of large countries such as China and India. Big or small, all surveys have problems which we must resolve.

Large surveys (both censuses and samples) involve large numbers of interviewers, which creates problems in training, supervision, and the collection and processing of great amounts of data. In a country such as India, for example, where many languages are spoken, interviewers must be found who can converse fluently in several languages.

Small surveys are usually samples. Because of their limited size, they can be relatively low in cost, and results can be obtained quickly. Market researchers and pollsters use them frequently, as most of us know, since we get mail and phone questionnaires all the time. But no matter how small the sample, if the survey is to avoid being a misuse, it must be designed and carried out properly. Unfortunately, we see many surveys that do not "follow the rules" of proper design and execution.

When a television program asks viewers to call in their responses to questions, it is conducting a kind of survey. The audience, which may not realize that such a "survey" is too unscientific to be valid, may incorrectly assume that a scientific process is taking place and that the reported results are accurate and precise determinations.

I Do Not Choose to Answer

In any kind of survey—a census or any of the many forms of sample surveys—the lack of a response from a portion of the surveyed group is a potential source of serious error and consequent misuse. If the *response rate*—the proportion of the surveyed group that responds—is reported with the results, you can draw your own conclusions on the value of the findings. But unfortunately, in the sampling process, the response rate may be confused with the proportion of the total population which is sampled (the *sampling proportion*), and the results may be applied to the population at large, a serious misuse.

We have seen survey reports in which the response rate was as high as 95% and others in which it was as low as 3%. A low *response rate* is a source of error because the nonrespondents may not respond for a reason that is related to the subject of the survey, making the results suspect (we discuss what is "low" later in the chapter). On the other hand, a small *sampling proportion* is usually of no significance. As Dr. George Gallup (the famous pollster who has released over 6000 polls in more than four decades of polling) points out, you do not need a large sampling proportion to do a good job, if you first stir the pot well:

> Whether you poll the United States or New York State or Baton Rouge (160,000 population), you need only the same number of interviews or samples. It's no mystery really—if a cook has two pots of soup on the stove, one far larger than the other, and thoroughly stirs them both, he doesn't have to take more spoonfuls from one than the other to sample the taste accurately [2, p. 11].

How Many Questions Should I Ask?

The number of questions asked in surveys varies from one to hundreds. A voting intention survey may ask no more than "Do you intend to vote?" But in an "in-depth" survey hundreds of questions are asked. The Archdiocese of New York surveyed the Hispanic population in its area to investigate 12 subjects of interest, including "Religious Identity," "Meaning of the Church," "Religious Education," and so forth. This in-depth survey contained 400 questions (3, p. 105 ff.). Customer product preference surveys fall between these extremes, asking for information concerning individual demographics (age, sex, family income, and so forth) and product preference (which product do you prefer and why?).

EXAMPLES OF SURVEY MISUSE

Verification: What Did They Say?

You can't always verify surveys in which people are asked questions. In most cases, the best you can do is make a spot check of some of the answers. This is what the U.S. Bureau of the Census does when it resurveys a sample of the original respondents for verification. We get some verification, also, when two polling organizations get substantially the same results from independent surveys.

Strict verifiability means that you must repeat the survey in the same way under the same conditions. For the kinds of surveys we are discussing, it is almost impossible to have the same conditions repeated, and unfortunately, strict verifiability is almost always impossible.

Even if you could reproduce identical conditions, repetition alone can change the individual's response. The U.S. Bureau of the Census data on reported length of unemployment is an example of this phenomenon. When making the monthly sample survey of employment and unemployment (Current Population Survey, or CPS), the bureau's interviewers question the same housing unit for four consecutive months and usually are able to reinterview the same persons. The interviewers ask those who say they are unemployed to give the number of weeks of unemployment, and this question is repeated as long as the same respondents say they are unemployed.

Since the time interval between interviews is known (four or five weeks), you would expect that the number of weeks of unemployment reported at the second interview would be greater than the first value reported by the number of weeks between interviews. Alas, in 1984, one study found that about three-quarters of the respondents gave inconsistent responses! Some "gained" weeks of unemployment between interviews, others "lost" weeks of unemployment (4). Which response was correct: the first, the second, the third, or none of them?

Economic factors often make verification impractical. One of us directed a survey of the employment of physically handicapped workers under a grant from the U.S. Office of Vocational Rehabilitation (5). After publication, another researcher wanted to investigate the same subject to confirm or refute the results. Since this survey required a sizable staff, he tried to obtain grant funding, without success. There was no verification.

Human attitudes and social environments change with time. Thus, the passage of time alone can prevent verification. If the employment researcher of the preceding paragraph were to obtain funds today, he could no longer verify the physically handicapped worker survey of 1959. Four decades later, he could only perform another survey which "duplicated" the 1959 survey in its questions and methodology. The outcome would be the basis for estimating the changes that have taken place since then, but strict verification would be impossible.

Lack of verifiability means that the reader often must accept or reject results based on the reader's confidence in the surveyor. Reputation counts. In the absence of a definitive way to verify survey results, your best defense is a thorough awareness of survey methodology, so that you can decide how to interpret the results of a particular survey.

Misuses in the Questions Asked

How many are the ways to incorrectly ask questions of respondents! In many cases, these defects may be the result of deliberate or inadvertent bias. There are many ways in which the same question can be asked (and answered), and it is not certain that all pollsters agree on the best way to ask questions. In addition, professional pollsters often disagree on which way of asking is most likely to produce valid responses.

There probably is no way to eliminate all problems from all questions in all surveys. We will discuss only two types of problems with questions: *leading questions* and *predetermined-answer categories*. Through these two examples, we illustrate the type of reasoning needed to deal with bias. The reader's best defense is to be aware of the nature and effect of bias.

Leading Questions—Getting the Answer You Want

When a question is worded so as to ensure getting a particular answer from the respondent, it is called a *leading question*. There is no shortage of examples of leading questions. Some years ago, one of us received a questionnaire from a member of the House of Representatives soliciting support for opposition to a bill limiting unionization. The second question of this questionnaire was:

> Do you feel that anyone should be forced to pay a union boss for permission to earn a living?
> YES _____ NO _____

It is extremely unlikely that anyone would answer "YES" to this question, for who wants to *force* someone to pay a union *boss* for permission to earn a living? But some people who do not want workers to be forced to pay union bosses for permission to work may think closed shops are good or that not all union leaders are bosses.

At a time when the Soviet Union was experiencing the economic and social problems that led to its dissolution, Peter S. Temes, then at Columbia University, commented on some not so subtle leading questions in a Soviet questionnaire about anti-Semitism (6). One of the questions asked whether the respondent agrees that "More than any other group in society, it is the Jews who are responsible for the problems the Soviet Union is experiencing today." In Temes' words, this question "validates the idea that distinct religious, ethnic or racial groups bear responsibility for the . . . problems of a multiethnic nation."

Another question asks if the respondent agrees that "when it comes to choosing between people and money, Jews will choose money." Temes feels that this question "validates the notion that all Jews act one way or another, [and] that there is something inherent about Jewishness that determines behavior." The questions are variations on the old saw, "Have you stopped hitting your spouse?"

Some leading questions about opinions reflect the thinking of the originator more than the opinions of the respondents. Here are several examples from the *1995 Legislative Questionnaire* of Representative Christopher Shays of Connecticut (7).

This comes from a section on the budget process:

G. Do you support balancing the federal budget in seven years?
1. Yes, we should balance the budget in seven years.
2. No, it is too short a period—we should take ten years or longer.
3. No, it is too long a period—we should take five years or less.
4. No, we don't need to balance the budget.
5. Other.

Apparently, whoever devised the question did not think that anyone would want to balance the budget in eight or nine years. Also, as questions 2 and 3 call for open-end (interval) responses* ("ten years or longer"), there is no possibility of a response except "Other" for those who might want the budget balanced in 12, 15, or some other number of years that fall in the category "ten years or longer." Since this does not involve a huge sample, perhaps it would have been better for Congressman Shays to ask the open-end, or free-answer, *question*: "How long should it take to balance the budget?" which offers the respondent an opportunity to give a more precise answer such as "11 years" or "three years."

Predetermined Answer Categories: Not Getting the Answers That You Don't Want

In many surveys, the respondent is given a list of answer categories from which to choose an answer. Usually, the respondent can't give any answer other than one which is specified by the given category.

Here is an example from House of Representatives member Christopher Shays's *1995 Legislative Questionnaire* (7):

> T. Most agree that our current welfare system is broken. Which step should we take first to reform it?

Six responses were offered to the recipients which discussed various measures to "reform" the "broken" system. They included times limits, paternity establishment, cracking down on illegal immigrants, and so forth. A seventh response was offered to those who were not among the "most" who agreed the system was broken:

> Welfare is not "broken;" we just need to spend more on the program.

This predetermined response leaves no answer for those who do not think that welfare is broken, but would keep spending constant or would decrease spending under the present program. This response makes it appear that anyone who does not think that welfare is broken wants to spend more

---

*An open-end (interval) response is one which has no upper or lower limit, such as "age 65 or over."

under the present program. There is an eighth response for "other," but that does not resolve this built-in twisting.

Open and Closed Responses

Howard Schuman and Jacqueline Scott of the University of Michigan's Survey Research Center studied the explicit effects of offering respondents fixed or open choices and the implicit effects of the wording of questions which allowed "open" responses (8).

First, they asked respondents answering the *open* questionnaire to answer this question (which offered no suggestions):

> What do you think is the most important problem facing this country today?

They then offered the second group of respondents answering the *fixed* questionnaire the same question, amended as follows:

> Which of the following do you think is the most important problem facing this country today in the following areas: the energy shortage, the quality of public schools, legalized abortion or pollution, or if you prefer, you may name a different problem as most important.

Of the respondents who answered the *open* questionnaire, 2.4% of the total number of respondents listed one of the four problems that were offered in the fixed questionnaire (which they had not seen) as follows: the energy shortage (0%); legalized abortion (0%); quality of public schools (1.2%); and pollution (1.2%). All other responses added up to 93%, and Don't Knows were 4.7%. These unprompted respondents were far more interested in unemployment (17%), general economic problems (17%), nuclear war (12%), and foreign affairs (10%).

But of those respondents answering the *fixed* questionnaire, a startling 60% chose (surprise!) energy, quality of public schools, legalized abortion, and pollution. This strongly suggests that what you offer the respondents, they will take. To what extent do such responses reflect the preferences of the respondents?

These researchers found that open questions had problems as well. In another experiment, they found that the wording of the open question could also constrain the respondents' replies. All questions must be open to scrutiny, and the results must be interpreted with great caution.

What Does It All Mean?

As discussed above, how questions are asked—and answered—influences the interpreted results. But the results quoted in the final presentation (which is

*"And now, Sirs, in your work—which do you find most
efficacious—soap flakes or scouring powder?"*

Figure 9.1 Some problems of how questions are worded and how samples are interviewed have been with us for a long time. The artist based this drawing on his
observations of polling practices in the 1930s. (Reproduced by permission of the estate
of George Mabie.)

often a newspaper article) may or may not truly reflect the opinions or intentions of the respondents. To evaluate survey results, you must know exactly
how a question was worded and how the sample was interviewed. In Figure
9.1 you can see how artist George Mabie views the surveyor's role.

Schuman and Scott (8) used an experiment to show how the form, or wording, of the question (open or closed) influenced the response. Good professional pollsters, rather than offering both types of question—open or closed—often request the same information using two differently worded questions in the same survey. This makes it imperative that not only respondents but those reading the results of a poll know exactly how the questions were worded.

In November 1997, *The New York Times* reported on two polls, taken before Election Day, which clearly illustrate how asking for the same information with two differently worded questions can give different results (9). The reason for polling was to try to predict the result of the public vote on a proposal to end affirmative action in the award of city contracts and in city hiring. To make this prediction, the Center for Public Policy at the University of Houston and the Baker Institute for Public Policy at Rice University jointly conducted two polls for *The Houston Chronicle* and TV station KHOU-TV, using the two versions proposed by two different groups.

In both polls, the respondents stated whether they were for, against, or not sure how they would vote for the proposal. In the first poll, the proposal was worded as follows:

> The City of Houston shall not discriminate against or grant preferential treatment to any individual or group on the basis of race, sex, ethnicity or national origin in the operation of public employment and public contracting [9].

We will call this the "nondiscrimination" wording. In the second poll, the proposal was worded this way:

> Shall the Charter of the City of Houston be amended to end the use of preferential treatment (affirmative action) in the operation of the City of Houston employment and contracting?

We call this the "affirmative action" wording.

Here are the results of the two polls*:

Wording	For	Against	Not sure	Other
"Nondiscrimination"	68.1%	16.1%	14.4%	1.40%
"Affirmative action"	47.5%	34.0%	17.9%	0.60%

---

*The sample was 831 registered voters; the 95% confidence interval has a width of approximately plus or minus 4%.

What this table shows is that the wording of the question produced dramatically different results, 68.1% for the "nondiscrimination" version versus 47.5% for the "affirmative action" version, both of which were supposed to accomplish the same political action: ending any kind of preferential treatment in city employment. If you try to decide whether the voters of Houston are for or against the proposal simply on the basis of the first two questions, it is difficult to arrive at a clear-cut answer. A headline based on only one of these two questions, saying either "Houston Voters Favor Ending Affirmative Action" of Houston Voters or "Houston voters Oppose Ending Affirmative Action" is a misuse of statistics.

Epilogue: The proposal, using the "affirmative action" wording, was put to the voters. The voters rejected the proposal by 55% against to 45% for the proposal. The poll predicted a vote of 47.5% for and only 34% against and 17.9% not sure. However, in an election, there are no "Not Sure" votes. Houston retained affirmative action. The "Not Sures" were the deciding votes against the proposal and for affirmative action in their city.

Sampling Done Here

We take samples because we cannot afford the money, time, or resources needed to query the target population. We want to project the results of the survey of a particular sample to its target population; but how many pitfalls there are on the way to those results!

A Depressing Survey

The year is 1936 and the surveyor wants to survey the population of all American voters to determine how they will vote. The size of this population is tens of millions, so a sample is taken. But from what listing of people are we to take our sample? It is easy to get lists of automobile registrants, telephone directories, and similar sources. But this was a time when, inconceivable as it may seem today, automobile and telephone ownership was a luxury. Thus, if the sample is taken from such lists, the sampled population is not the target population (of all voters), and this sample is probably of a relatively high economic status. We should not be surprised when the sample is wrong about the voting behavior of the voters, as was the survey based on the sample just described, taken by the *Literary Digest* in 1936 (10). The magazine said that Alfred M. Landon would win the election. Instead, Franklin D. Roosevelt was elected with 61% of the popular votes cast. This error was directly responsible for the death of the magazine, which is now remembered only for its notable misuse of sampling.

Self-Chosen People

A surveyor can take extensive precautions to assure that the sampled population is the same as the population of interest, but if it is possible for the recipient of a questionnaire not to respond, the results can be seriously distorted. For example, the *Literary Digest* survey used mail questionnaires. Only 20% of the recipients responded—the *self-chosen*.

> There is general agreement that this mail ballot method was subject to a serious distortion because the better educated and more literate part of the population, as well as those who were higher on the economic scale, tended to return their ballots in greater proportion than those who were lower in educational and economic status [10].

Misusers continue to make projections from the self-chosen to the whole population, which rarely gives a correct answer and creates a gross misuse.

Many of the poll results appearing in magazines and newspapers fall into this category. Readers are invited to send in their responses to published questionnaires. A modest fraction of the readers respond, and these responses are tabulated and usually allowed to stand as representing the readers' opinions. It is a gross misuse to project such results to the population as a whole, but headline writers have a field day with these surveys. Such cautions as are given are often superficial and misleading, as for example, this gem from *Modern Maturity* in discussing the results of a published questionnaire titled "Who Wants a Second Career?":

> What came out of our computer cannot speak for people over 50 in general because those who don't want to work didn't write in. . . . But the feelings that moved you to respond are a fair sample of what's on the minds of you who do want to work in your later years [11, p. 31].

Modern Maturity reported that 4136 readers responded to the magazine's request for information about those who are interested in second careers. Fair sample? Around 4000 *self-selected* replies out of 30 million subscribers! As to the statement that "those who don't want to work didn't write in," the author immediately goes on to report several negative comments from respondents who did *not* want to work. As we have noted, misuses tend to come in clusters.

Technological developments continually give new ways to carry out old misuses. The two-way connection possible with certain cable TV installations allows the use of "instant polls" in which viewers can, *if they choose,* push buttons and answer questions presented on the video screen. The results are tabulated within milliseconds and summarized on the same TV screens. You can't fault the technology.

But you have good reason to be skeptical of results drawn from such polls, whether or not they use sophisticated technology. To get a feeling for the kind of things you have to think about when viewing such poll results, let's review how a response comes about. To become a respondent, there is a hierarchy of self-selection. The potential respondent must have installed the necessary equipment (cable, television, response device, or telephone), selected the particular channel and then the particular program asking for responses, and then felt that the issue was important enough to them to make the response (which, in the case of self-respondent telephone polls, will require a payment).

These are valid data in the sense that they represent the responses of those who choose themselves through the hierarchical process described above.

There is nothing inherently wrong in such a self-selected sample when the reported results make it clear the nature of the response process. However, it is a major misuse of statistics when a Dr. Nowall projects the results of a survey of such a sample to a population that is not represented by the sample.

For example, the ABC network, reporting on the 1980 Reagan–Carter presidential debates, "made no claim that the method of polling would be a good predictor, and during the following day, indicated that there was some controversy as to the scientific accuracy of the method" (12). It would have been more correct to say that, of those who watched the presidential debates and chose to spend 50¢ to respond, the opinion was 2:1 in favor of Reagan. This is a valid statement, but to project this result to all listeners or all voters is a misuse of statistics.

The same statement can be made in regard to the *Modern Maturity* poll or any other similar poll relying on the self-chosen. Our many examples show that we cannot always rely on the public, the headline writers, or policy makers to pay close attention to "scientific" qualifications which would limit the conclusions only to the sample. Unfortunately, public opinion and national policy are frequently influenced by invalid projections of such sample survey results to the whole population.

For example, here is a congressman relying on self-chosen respondents for aid in making legislative decisions:

> Congressman Hollenbeck has received from the survey an extremely valuable information source to aid in making legislative decisions best representing the views and concerns of [N.J.] Ninth Congressional District residents [13].

Each year, Congressman Hollenbeck mailed questionnaires to voters in his district and requested replies. In the report cited above, he notes that he received 16,000 replies. Because the respondents were only those of his constituents who felt strongly enough about some issues to respond, and took the time to respond and mail back the questionnaires, the results cannot be statistically projected to the population of all constituents as might be done with a random sample. Only if Congressman Hollenbeck had made a scientifically chosen random sample of his constituents could he claim to know how *all* his constituents felt about particular legislation.

The Good and the Bad: Telephone Interviewing

In the beginning, interviewers sought out respondents and dealt with them face to face. If the sample of respondents was scientifically chosen, if all who were approached responded, if the questions were well phrased and the interviewers were trained and skilled, the quality of data collection was assured.

Later, someone had the idea of sending questionnaires by mail, and questionnaires were sent to the four winds even as pollen is distributed by air currents. The analogy is not so far-fetched for, like pollen, not all the questionnaires found anyone home and not all that settled down received attention, making the problem of the self-chosen respondent important.

Today, we have the telephone interview. Surveyors have found that the telephone interview is less expensive even when, in a few cases, the surveyor must send an interviewer for a face-to-face interview. Studies have shown that the results of telephone interviewing differ little from face-to-face interviewing for some subject matters (14). In such cases, any decisions based on the survey findings would be essentially the same no matter how the interviewing was done.

However, many poor persons—often members of a minority group which may be the target population—have no telephone in their homes. In 1994, 6% of all households had no phones; 13% of those with annual family incomes below $10,000 had no phones. In minority households, 14% of all black and Hispanic households had no phones, and about 18% of black and Hispanic households with annual household incomes below $10,000 had no phones (1, Table 883).

The proportion of households of all kinds without phones has been declining, but even so, the minorities and the poor and unemployed are underrepresented among phone owners. People without phones cannot be reached by a telephone survey and will not be represented in such a phone survey. Suppose Dr. Nowall is making a telephone survey of urban housing problems of the poor or minorities. Can he project the results to include all the members of those categories?

Thus, the purpose of the survey can determine the accuracy of a telephone sample, because it is the survey's purpose which delineates the target population. If the purpose of the survey calls for a target population of middle- and upper-income families, there is no problem with a telephone survey. Suppose, for example, that a market researcher wants determine the attitudes of potential buyers about fully air-conditioned homes. A projection to the target population will not be invalidated by phone ownership characteristics because all the households likely to be interested customers have phones. But in a survey such as the Archdiocese of New York survey of the Hispanic population (3), the conclusions from a telephone survey could have been disastrously flawed.

Not Everyone Will Talk to You

A sample may be designed to be appropriately random so that probabilistic projections could be made, but design is one thing, implementation another. The nonrespondent is the major problem in implementation. If, for example, only 10% of the chosen sample responds, probabilistic analysis cannot be used (standard errors are without meaning) and the results may be totally useless. Why do some people refuse to be interviewed? If the reasons for refusing to be interviewed are linked to some characteristic of the survey questionnaire, then the results may be biased. Thus, in the implementation of the survey, every effort must be made to get 100% response rates from the chosen sample.

What response rate is adequate? Anything less than 100% is suspect. Good surveyors aggressively and skillfully try to get 100% response. If the interviewing is done face to face or by telephone, it may be possible to make return visits or calls. The problem of obtaining responses is more complicated in mail surveys. Professional surveyors use many procedures for dealing with the failure to respond. Without going into the details of these procedures, we advise readers to be wary of mail surveys. And if the report doesn't tell what proportion of the sample responded, *beware.*

Even when the proportion is reported, it may be unsatisfactory. A survey to determine attitudes of both business people and academics toward several aspects of master of business administration (MBA) degree programs included both the CEOs and personnel directors of many companies. The results were that:

> Business participation was low, even though the presidents' and personnel directors' response rates fell within the 20% to 25% range that is standard for mail surveys. Readers should be aware, therefore, that the business response is not as representative as we had hoped it would be [15].

The "20% to 25%" who responded were self-chosen. The disclaimer quoted above is a mild caution where a drastic warning is needed because the results of this survey (based on a response that represents the opinions of a self-chosen minority) could be used to determine educational policy for MBA programs.

Cost limits the amount of follow-up and, thus, the response rate. Usually, even with intensive efforts, it is almost impossible to get 100% response (deaths, disappearances, physical incapacity). If we have reason to believe that the nonrespondents are a random sample drawn from the same population as the respondents, then the two groups have the same characteristics, subject to chance effects. Is there any way to come to this conclusion at some reasonable level of confidence? One way is to take a modest-size random sample from the nonrespondents and then concentrate extraordinary efforts on getting responses. This "survey within a survey" can be used to make confidence intervals on the characteristics of the nonrespondents.

In the absence of 100% response, or information on the nonrespondents, the probabilistic computations (confidence intervals, levels of significance) are in error. The larger the proportion of nonrespondents, the worse the error. Even at a nonresponse level of 50%, probabilistic statements are likely to be in serious error. The good surveyor determines how the nonrespondents might affect the observed results and establishes their characteristics.

To the best of our knowledge, a response rate of 100% is rarely achieved. Then what response is acceptable? It is not possible to give an exact value for the minimum acceptable response rate and to reject any survey with a response rate below that value. We can say that the closer we are to 100%, the more confidence we can have in the findings.

Government surveys often have a high response rate because they are subject to public oversight, have relatively generous resources, are often continuing surveys, and conducted by professionals with a commitment to statistical analysis. For example, the U.S. Bureau of the Census' monthly sample survey of the U.S. population had response rates near 95% (16, p. 197).

Unfortunately, some survey reports fail to even mention the response rate. The report of a professional polling organization discussing registered Democrats says only:

> A total of 850 registered Democrats in New York State were interviewed by telephone in late April. These respondents were randomly selected from current voter registration lists in 100 Election Districts. The 100 Election Districts were selected by a random method according to which an Election District's chance of being selected was equal to its proportion of the total number of combined votes cast in the 1976 U.S. Senate and 1978 Gubernatorial Primaries (17).

This description, with its careful (and correct, for these purposes) method of choosing the sample, cannot be taken to mean that the actual sample obtained and used was valid. We do not know what proportion of the sample originally designated is represented by those 850 respondents. Our suspicion that the response rate is low is heightened by the fact that 850 registrants out of 100 districts is an average of eight registrants per district. Surely at this low rate, some of the districts would have had zero respondents.

Aging in America: Trials and Triumphs (18) is based on a telephone survey of the U.S. population age 60 and over. The purpose of the study was "to identify the negative forces in society that older Americans perceive hurt them the most—and the positive forces that they perceive provide their most effective support systems" (19, p. 2).

The first summary finding is that "67% of senior citizens say they always feel useful; 65% report strong self-images; 61% feel that, in general, things are worthwhile; 56% are serene; 52% show a high degree of optimism" (19, p. 13).

These numbers—67, 65, 61, 56, 52—imply accuracy and precision. We can evaluate them because this report gives some information on the responses. What do we find in regard to the responses? The survey reached 902 individuals age 60 and over; how many in the original sample design were not "reached" we do not know. Of these 902 contacts, only 514 supplied whole or partial data. Only 481 respondents supplied full data! Assuming that the original design called for a sample of 902, only 53% responded.

And what characteristics might the non- or partial respondents have in common? Could they be the individuals with insufficient self-image to participate in a phone interview or to complete one? Could they be the physically disabled who feel useless? Could they be the mentally agitated who are not serene? Could they be the depressed who have neither optimism nor the capacity to deal with a stranger interviewing them about their feelings on the telephone?

Individuals over age 60 living in institutions were not contacted, nor were those without telephones. This occurred either from poor sample design or the effort to hold down costs. The bias is in the direction of a more positive result than if the whole population were sampled and the nonrespondents evaluated. Older people who are living in institutions are more likely to be ill or disabled and less likely to be as optimistic as the survey report portrays older people. As we discussed earlier, poorer people are less likely to have telephones and we suspect they are not as optimistic as the portrait given in *Aging in America*.

Furthermore, no comparisons were made with the population under 60 years of age. We need that information to make a useful interpretation of the survey results. Misuses travel in packs.

In 1991, a scientific panel and the General Accounting Office concluded that the Department of Agriculture's 1987–88 National Food Consumption Survey was so seriously flawed as to be almost useless:

> The worst problem is the survey's low response rate of 34%, making it questionable whether the data are representative of the population, the G.A.O. said. Such surveys also require follow-up surveys of those who do not respond. National Analysts, which conducted the survey, told both

the Agriculture Department and the General Accounting Office that its
follow-up data had been lost, but finally acknowledged that no follow-
ups were conducted . . . [20].

What excuse can they have for such low response rates in a survey costing $8
million? Certainly not that it can't be done. The Department of Health and
Human Services runs a similar survey which gets response rates of from 77%
to 86%. Even with this criticism, we find that the results from this flawed
survey are being tabulated in the Statistical Abstract of the U.S. 10 years later
(20). Give National Analysts "A" for knowing that follow-up was required,
and "F" for integrity.

MORE TO COME

We close the first part of our discussion of surveys and polls here. We will
summarize both parts of our discussion in Chapters 9 and 10 at the end of
Chapter 10.

10
Surveys and Polls: Analyzing the Data

The person who must have certitude, who cannot embrace conclusions tentatively, should not be engaged in social scientific research.
—Norval D. Glenn

Well, I'll be damned if I'll defend to the death your right to say something that's statistically incorrect.
—Character in a cartoon in *The New Yorker*

Every number is guilty unless proved innocent.
—Anon

INTRODUCTION

In Chapter 9, we looked at some of the ways in which the basic data for a survey can be collected, and the misuses that can arise in that collection process. In this chapter, we discuss some of the factors in the analysis of survey data. Our emphasis is on empowering you to do enough analysis yourself that you can verify the results of published surveys.

All survey results are uncertain, to a greater or lesser degree. Statisticians call this uncertainty "error" (this "error" does not necessarily refer to a mistake). Statisticians often break down such error into a cascading series of contributing errors. If the sample was chosen randomly, a statistician can scientifically calculate the magnitude of the contribution to error due to the sampling process itself, which is called "sampling error."

The contribution to total error which is not due to sampling is usually called, logically enough, "nonsampling error." The term "bias" is also used for this type of error. There are many forms of bias that enter into the nonsampling error. It is rarely possible to be exact about the magnitude of biases, but they must be taken into account, and it is our intent to give you a basis for doing so. (Of course, a nonsampling error might be a true error, such as incorrect measurements, mistakes in data entry, and so forth.)

Fortunately, even the statistically unsophisticated reader can evaluate the results and confirm or refute reported results.

UNCERTAINTY

What Is Uncertainty?

By its very nature, a sample cannot guarantee results that are the same as the results of a census of all the individuals in the population of interest. For example, try to determine something as simple as the proportion of males in some population. The sample may give a proportion that is exactly the same as in the population (unlikely), near the value (likely under the right circumstances), or very far from the value (unlikely). This example is expanded below.

This uncertainty—which is inherent in the random process of choosing a sample—we call *error, variability,* or *uncertainty.* If you choose the sample of individuals to be queried by some probability rule in a suitably random manner, then the sampling error is your estimate of the precision of the sample process which gave you the results. In this manner, the uncertainty in the results due to the sampling process can be expressed as a "margin of error," plus or minus some amount.

Precision talks! *Precision* tells how well the measurement has been performed. Let's use the simple measurement of length as an example of precision. Then we will relate it to sampling error. The meter is defined as 1,650,763.73 wavelengths of red-orange light given off by krypton-86 under certain specified conditions. If you mark this distance off on your meter stick to represent 1 meter, then your meter stick is highly *accurate.* If this meter stick has millimeter markings on it and you use it to measure length, any measurement will be to plus or minus 1 millimeter; this is the *precision* of your measurement. If it had only centimeter markings, any measurement would be to plus or minus 1 centimeter, and the measurements would be less *precise* even if the meter stick has millimeter markings and, thus, is precise to 1 millimeter; you cannot be sure of its *accuracy* unless you checked it as described above, or compared it with a reliable "standard" which was checked in that way.

In surveys, precision is a function of the sampling process, not the markings on a scale. Survey precision is usually measured in terms of a numerical interval higher than, or lower than, the reported result, analogous to the "plus or minus" 1 millimeter in our meter stick example. A survey result may report an estimate of proportion as "40%, with a margin of error of five percentage points." Thus, the prior statement of precision means that the estimated proportion is from 35% to 45%. If the author of the report—researcher, analyst, journalist—has observed good practice, the published results will include a value for the margin of error.

Precision of a Survey When Estimating a Proportion

If the sampling process is complex, it may take a statistician to calculate the precision of the estimated value. However, in many cases, even a nonstatistician may be able to determine the precision. A good survey report gives enough information to verify the reported precision. When sufficient information is *not* given, the reader—statistician or not—is prevented from knowing or calculating the precision, and the reader cannot evaluate the reported results.

As mentioned before, when we get the precision of a sample survey we use a margin of error, or "plus-or-minus" concept, just as we do for measurements of distance, weight, and other physical quantities. Consider making a survey of a given population to determine the proportion of males, and assume that the proportion in the population is one-half, .50. If you took a great many samples by a scientifically determined method, you would certainly get many different sample proportions, such as .40, .49, .51, .54, and so forth, because of chance variation.* Of course, you never do take a series of identical samples; you take only one. How do you evaluate the precision of that one sample?

We take our one result and put a "plus-or-minus" interval around it. It is customary in survey sampling to choose this interval for a "95% confidence level." This means that, if we repeatedly took the same sample and made a plus-or-minus interval in the same way, 95% of the time our interval would contain the true value. Thus, if we interviewed 100 respondents and got a result of .40 for our sample of the population for which the true value of the proportion of the surveyed characteristic was .50, we have an interval of .30 to .50.

---

*You can make your own experiment. Toss 10 coins (or more if you have the patience), count the number of heads, and compute the proportion of heads. Since the probability of getting a head is one-half, this is analogous to taking one sample from a population in which the proportion of heads is one-half. Repeat the toss several times. Each time you make a toss you can compute the proportion of heads. They will not all be one-half! Note the mean and spread of the proportions. From this you can get an idea of the precision of the sampling process. Try it again with 40 coins, and note once again the mean and spread of the proportions. This is a good way to get an appreciation of the nature of precision in sample surveys.

If you make plus or minus two standard error intervals around the survey result, then you will have an interval corresponding to a 95% confidence level. This, as you will see, is what is customarily reported as the margin of error in those surveys (so often seen) that estimate proportions, such as voting preferences, opinions, and the like.

This is a hypothetical situation, since if we knew the true value we would not need to make a survey. The value of considering this hypothetical situation is that it is the basis for our evaluation of the process of surveying a population *once,* and evaluating the precision of the result of that survey.

How did we arrive at the conclusion that the interval for the estimate is from .30 to .50? It's easy. The unit of measurement of the variability of the result is the *standard error,* which depends primarily on the sample size. For samples taken in surveys where the sample size is a small fraction of the population size, the *fraction of the population taken* in the sample is unimportant.* What *is* important is the *sample size.*

How do you calculate the standard error? For most simple surveys involving proportions, it is remarkably uncomplicated. The standard error is:

$$\sqrt{\frac{(\text{Observed proportion}) \times [1 - (\text{Observed proportion})]}{(\text{sample size})}}$$

If you did not know the true proportion of males in a population, took a random sample of 100 persons from a population to determine the proportion of males, and got a sample proportion of .40, the standard error would be:

$$\sqrt{\frac{(.40)(.60)}{100}} = \sqrt{.0024} = 0.05$$

The width of the 95% confidence interval is:

$$\pm 2 \times .05 = \pm.10$$

Thus the 95% confidence interval is .40 plus or minus .10, or .30 to .50, or 40% plus or minus 10%. If this value were reported to the public, the margin of error would be reported at 10%. This apparently poor precision is the consequence of the small sample size. The surveyor must decide, before accepting these results, whether this precision is satisfactory.

---

*When the sample size is a "large" fraction of the population size, the standard error described by us is multiplied by the *finite correction factor,* which is less than 1. How large is "large"? If the sample size is one one-hundredth of the population size, then the finite correction factor is .995. If the sample size is one-tenth of the population size, then the finite correction factor is .95; when the sample size is one-half of the population size, then the finite correction factor is .70. How often do you see surveys in which the sample size is more than one-tenth of the population size?

This simple calculation shows you how a misuse can be created in determining an estimate of a proportion. *If neither the standard error, the margin of error, nor the sample size is given, you cannot know the precision of the result.* What is usually reported to the public is the sample size and the margin of error. However, in reports on surveys involving complex probability designs, such as those carried out by governments, market research firms, and many scientific investigators, you will find the standard error explicitly stated for each resulting value.

To measure sampling variability in the way we describe, the sample must be a "random sample" where every member of the population is given an equal chance of being chosen.* If the sample was self-chosen, haphazard, or directed by the judgment of the surveyor, you cannot say that the 95% confidence interval is the observed value plus or minus 2 standard errors. Beware the overconfident pollster!

The Precision of a Survey That Estimates a Mean

If the result is an average (the arithmetic mean) and not a proportion, the computation is similar. The 95% confidence interval is plus or minus 2 standard errors on either side of the sample mean. But you need to know more than the sample size and the observed sample mean. You must also know, or be able to estimate, the standard deviation of the observed variable in the population. If the specific data values obtained in the survey are available, you can estimate the standard deviation by a computation that can be made on a pocket calculator. You can then get your estimate of the standard error by dividing the standard deviation by the square root of the sample size. A good survey report will give either the standard deviation or the standard error if the original data are not given.

How simple are these principles! The smaller the standard deviation of the variable, the smaller the standard error; the larger the sample size, the smaller the standard error. How widespread is ignorance of these principles! Here is an example of ignorance on the part of people who should know better.

In 1997, "luring" advertisers to the Internet (and by implication, away from TV) was a major concern of America Online (AOL). AOL commissioned Nielsen Media Research to determine whether their subscribers watched TV less than nonsubscribers (1). One way to make this comparison is to determine an estimate of the mean weekly TV watching time of both AOL subscribers and nonsubscribers. Nielsen surveyed 262 U.S. households that were AOL subscribers and 4982 that were not AOL subscribers. The sample sizes are known, and all that is needed to determine the margin of error is the stan-

---

*This is a simple definition which is adequate for the purposes of our discussion.

dard deviation of the viewing hours for the two categories of survey subjects. Since the standard deviation has a direct effect on margin of error for an estimate of a mean, knowledge of sample size alone is not sufficient, as it is in the case of an estimate of proportion.

The survey found that the mean TV watching time of AOL subscribers was 15% less than that of nonsubscribers. Possible good news for AOL and possible bad news for TV networks. The senior vice-president of research for the TV network Fox Broadcasting noted that the 262 homes used to draw the conclusions in the survey were "an awfully small number of homes." "The margin of error on that size sample would be exceedingly high," he said (1).

A spokeswoman for Nielsen said that "not all the data in the survey had been analyzed, so no margin of error could yet be determined." But she agreed that the margin could be "high." Yes, it *could* be high if the standard deviation of viewing hours was high, and apparently the analysts have not yet computed a value for the standard deviation. But until they compute and report that value, no one can say whether or not the number of AOL subscribers (262 in this case) leads to an exceedingly high margin of error.

Even if the margin of error turns out to be low, these results do not show that viewers are leaving TV for the action on the Internet. The same survey showed that the AOL subscribers had higher incomes and more education than the national average. (It would have been nice to know how they compared to the 4982 nonsubscribers in the survey, but this information was not reported in the *Times* article.) It's possible the members of this group may have watched less TV *before* becoming AOL subscribers.

USING SAMPLING ERROR WHEN EVALUATING SURVEY REPORTS

Below are four examples which illustrate some of the issues involved when evaluating survey results using sampling errors.

Family Income

Good surveyors evaluate precision or give their readers the means to make their own evaluation. For example, the U.S. Bureau of the Census published data from its March 1995 sample survey (the CPS) which included both the mean and median incomes and the standard errors for the means (2, Table 5).*

---

*As discussed in Chapter 7, the median income is the "middle value," that value which splits the data into two equal parts. For all practical purposes, half the families have less than this value, half have more than this value. Variability can be calculated for the median as well as the mean, but we have not discussed how this is done.

Table 10.1 Mean Family Income, Standard Error, and 95% Confidence Limits

	Bureau of the Census survey results		95% Confidence interval limits	
Race	Mean income	Standard error	Lower limit	Upper limit
White	$53,596	$340	$52,916	$54,276
Black	$34,011	$698	$32,615	$35,407
Hispanic	$32,654	$756	$31,142	$34,166

The width of the 95% confidence interval for the mean is plus or minus 2 standard deviations, an easy calculation. For example, the 95% confidence interval for the mean income of white households is from $53,596 plus or minus twice $340, or $52,916 to $54,276.

Table 10.1 shows the sample mean incomes and standard errors for the three major racial groups, using these statistics. As you can see from this table, the confidence intervals for black and Hispanic mean family incomes overlap. Neither the black nor Hispanic confidence intervals come anywhere near overlapping with the interval for white families. Thus, you can reasonably conclude that white household mean income is greater than the Hispanic and greater than the black, which in turn are essentially the same value.

If the sampling variability had been so great that all the intervals overlapped, even though the sample means were different, your conclusion would be different. In that case you would have to conclude that the survey gives you no basis for saying that there are differences in mean household income among these three groups.

To say that "in 1995 the income of the average white household is $53,596" is an incorrect statement—a type of misuse of statistics that we often see. The correct statement is that, based on our survey, we are "95% confident that the mean income of *all* white households is between $52,916 and $54,276."* Since stating these figures to the single dollar implies a precision that doesn't exist (called *spurious precision*), it is better (and easier to grasp) to say "between about $53,000 and $54,000."

Unemployment

A *New York Times* article reported that "some states scored heartening declines in joblessness; New York's unemployment rate was 6.8% compared with 7.4% in October" (3). The "heartening decline" is .6%, less than half of

---

*If this survey were repeated an infinite number of times (an obvious impossibility), the interval calculated this way would contain the true mean of all white households 95% of the time.

the variability of 1.5%, which is 2 standard errors (4). The decline of .6% is meaningless in light of the sampling variability, not "heartening."

Headline writers for U.S. government publications are not necessarily more careful than headline writers for newspapers. The U.S. Bureau of Labor Statistics issued a report (May 6, 1983) with the headline, "Unemployment Rate for New York City Declines to 9.2% in April." The reported decline was 1.8%, but the 95% confidence interval width for the monthly changes at this time was more than plus or minus 2.5%. The change between March and April may well be only a chance effect due to sampling.

A Survey Report in Which We Have No Confidence: When No Information Is Given

Given this rule, what are we to make of the 1997 story published in *Physicians Financial News,* titled "Have we lost confidence in our healthcare system? 80% say yes" (5)? Nowhere in the article is either the margin of error or the sample size given. Do we have any confidence in the results of the reported survey? One hundred percent of your authors say no.

A Reporter in Whom We Have No Confidence: When Incorrect Information Is Given

In 1997, the *San Francisco Chronicle* reported on the results of a poll to determine the popularity of Governor Pete Wilson of California. The report of the pollsters, the Field Institute, was shown in a table and followed our principles of reporting: "The poll was conducted February 11–17 by the Field Institute. . . . Governor Wilson's ratings are based on a sample of 500 adults. The overall margin of error is plus or minus 4.5 percentage points" (6).

Fair enough, and you can verify yourself that for the governor's approximately 50% approval rating and the sample of 500, the margin of error, based on a 95% confidence interval, is about plus or minus 4.5%. But the reporter seems to have been bewildered by the meaning of the margin of error: "Field said that 95 percent of the time, such a survey would have a sampling error of plus or minus 4.5 percentage points" (6).

In view of Field's other correct statements about sampling error, we doubt that they would have made such a statement. The worst part of this misreporting is that some members of the reading public will have taken away this bizarre concept as a definition of margin of error. They deserve better from the press.

Is It Always 95%?

So far we have made all our confidence intervals with a 95% confidence level. This is the level of confidence used to define the confidence intervals in much scientific work, as well as surveys reported in the media. You are not obligated to use this level; if you can live with a lower confidence level, then you will have a narrower interval.

As we discussed above, the Bureau of the Census reported that in 1995, the mean income for white households was $53,596 and the standard error $340. The 68% confidence interval will have limits that are the sample mean plus or minus 1 standard error. Thus, we can compute the 68% confidence interval for the mean white household income in 1995 as $53,596 plus or minus 1 standard error, $340, or $53,256 to 53,936. Remember, the confidence level (here, 68%) is the probability of the truth of your statement that the confidence interval includes the mean of the population from which the sample is drawn. The narrower the range, the lower the confidence level and the lower the reader's confidence that the reported interval contains the population mean.

Do you require even more confidence than you get from a 95% confidence interval? Then you can use a 99.9% confidence interval: Mean white household income is from $52,576 to $54,616, since the width of the 99.9% confidence interval is plus or minus 3 standard errors ($1020). Thus, the wider the range, the higher the confidence level and the higher the reader's confidence that the reported interval contains the population mean.

ON THE BIAS: NONSAMPLING ERRORS

How many ways there are to go astray! We could list all the sources of such error that we know, and the list would be incomplete by the time you read it. New situations, new technologies, and the like, all conspire to contribute to nonsampling error. Below we categorize and name sources of error which arise frequently. Categories are slippery, however, and here we offer three other categories in addition to those given in Chapter 9, in the hope of inspiring you to stay vigilant.

Interviewer Bias

The category of *interviewer bias* includes those instances where the nature of the interviewer or the process creates an error. For many years, statistics teachers have used, as an example of how bias can affect survey results, the example of a World War II Army survey of black soldiers which was conducted to determine their level of satisfaction. When white interviewers were used, the level of satisfaction was significantly higher than when black interviewers were used.[*] In general, social scientists and statisticians will automati-

---

[*]For all we know, this story may be apocryphal. We heard it many years ago, but no longer have any bibliographic reference. However, it seems reasonable enough.

cally assume that in any situation where ethnic identify is important, it is best to have an interviewer of the same ethnicity as the survey subjects. However, one of us is familiar with instances where the subjects will not communicate honestly (and in some cases at all) with interviewers of the same ethnicity or from their geographical region because of cultural factors. For example, in many cultures, admitting to rape not only exposes a woman to humiliation, but may prevent her subsequent marriage. Such a woman will not—and we know, often does not—admit rape to someone from the same culture, especially if it is possible that the interviewer has connections in the same region. This is the other side of the bias coin from the World War II example, and it illustrates the danger of making universal generalizations in designing surveys without specific knowledge of what will work in different cultural settings.

Interviewer bias occurs in less threatening situations as well. When teams from different administrative units are collecting data on the same questions, a "house bias" can result. This is the result of different training and supervision practices. This bias is often evident in different proportions of "Don't Know" responses when the same questions are asked by different teams. Some teams will have been trained and directed to aggressively try to get a substantive response other than "Don't Know," others will not be as diligent (7).

Wording Bias

There are no easy rules to avoid bias due to misunderstandings or cultural differences in the meanings of words. Some words have emotional meanings that transcend logic. As Michael Kagay and Janet Elder point out:

> A poll finding can turn on the connotations of a single word. A classic example is a pair of questions asked for years by the National Opinion Research Center at the University of Chicago and recently by the New York Times/CBS News Poll. The public was asked if the United States was spending too much, too little or about the right amount of money on "assistance to the poor." Two-thirds of the respondents said that the country was spending too little. But when the word "welfare" was substituted for "assistance to the poor," sentiments swung the other way. Nearly half the respondents said that too much was being spent on welfare [8].

Sponsor Bias

In public policy debates, the results of surveys are frequently quoted to support or attack a position. Creative and counterintuitive approaches can have great value. In July 1997, Professor John R. Lott Jr., of the University of Chicago School of Law, made the claim, based on his study, that arming more citizens with hand guns reduces crime (9). We do not concern ourselves with the methodology of the survey, which has been criticized elsewhere. We do think that interested hearers might well ask if what we call *sponsor bias* is at work in affecting the conclusions and perhaps the work itself. Professor Lott is the John M. Olin Fellow at the University of Chicago, and thus obtains part or all of his income from the Olin Foundation, which, in turn, receives much of its support from the Olin Corporation (10). The Olin Corporation manufactures Winchester ammunition. Does this mean that the professor produced a biased study? No, but it does raise our suspicions about the conduct of the study. Even without deliberate intent, the researchers could well have posed leading questions, or produced any of the many biases that can result when starting with an advocacy point of view. Should we should reject the professor's study out of hand? No, we must carefully evaluate the study, in view of our knowledge of Professor's Lott's source of funding, before accepting its conclusions as valid and reliable.

PAY ATTENTION TO THE PROFESSIONAL POLLSTERS

Professional pollsters have an excellent reputation for designing surveys and describing sources of errors in their statistics that could lead to misuses. They are often leaders in conducting tests to reduce the number of such errors and innovative in finding new approaches. While the totally error-free survey may always elude us, professional pollsters often succeed in reducing errors to tolerable levels and giving the reader sufficient information to evaluate the results.

We don't expect that most of our readers, unless they are professional pollsters, will design and carry out a survey, but the knowledge of good practice from the viewpoint of a professional can help in two ways: (a) in making an informed judgment of the proposals and plans of those who will carry out the work, especially if you plan to use the survey to get information; and (b) by increasing the knowledge base for evaluating surveys.

Some Good Advice

From Government Pros

Professional pollsters have publicly given much good advice, which, if taken by their clients, the public, and others, would greatly reduce misuses of statistics. For example, on nonsampling errors (bias, those involving accuracy) the U.S. Bureau of Labor Statistics tells us:

> Nonsampling errors in surveys can be attributed to many sources, e.g.,
> inability to obtain information about all cases in the sample, definitional

difficulties, differences in the interpretation of questions, inability or un-
willingness of respondents to provide correct information, inability to
recall information, errors made in collection such as in recording or cod-
ing the data, errors made in processing the data, errors made in estimating values for missing data, and failure to represent all sample households and all persons within sample households [11, p. 198].

From a Business Pro

Daniel Yankelovich, former president of a public opinion research organization, warned readers about an aspect of political polls we have not discussed:

> As opinion polls have grown more influential in recent years, the public
> needs to gain a better understanding of a seeming paradox: The polls are
> almost always accurate in the narrow sense of reporting what cross-sec-
> tions of Americans actually say in response to particular questions at a
> given time. Unfortunately, though, even accurate polls can be mislead-
> ing because what the people say is often not what they really mean.
>
> There is nothing mysterious or unsatisfactory about this. It is not a
> technical problem having to do with sampling, the phrasing of questions,
> or the tabulation of statistics. Nor is it a moral problem. People almost
> never lie outright in polls and they virtually never seek to mislead. When
> faced with the eventuality of an important decision, most people do not
> sort out their convictions until they have spent weeks or months "work-
> ing through" their feelings and attitudes. . . . at any point along the way
> a public-opinion poll may catch an attitudinal "snapshot" of the public
> in the act of making up its mind [12].

From Your Authors

We think that all media—print and electronic alike—should adhere to a minimum standard for reporting on their polls. *The New York Times*/CBS polls set
a reasonable minimum standard. They include a box with all poll reports which
gives the major information needed to evaluate the sampling process. Here is
a recent example:

> The New York Times/CBS New Poll on women's health issues is based
> on telephone interviews conducted May 19 through May 23 with 1,453
> adults across the United States.
>
> The sample of telephone exchanges called was selected randomly by
> a computer from a complete list of more than 36,000 active residential
> exchanges nationwide. Within each exchange, random digits were added
> to form a full telephone number, thus allowing access to listed and un-
> listed numbers. Within each household, one adult was designated ran-
> domly to be the survey respondent.
>
> The results have been weighted to take account of the household size
> and number of telephone lines into the residence, and to adjust for varia-

tions in the sample relating to geographic region, race, sex, age and education. In theory, in 19 out of 20 cases, the results based on such samples will differ by no more than three percentage points in either direction from what would have been obtained by seeking out all American adults.

For smaller subgroups of people the potential sampling error is larger. The sampling error for either men or women taken as a group, for example, is plus or minus four percentage points.

To insure that there were [sic] a sufficient number of black women, they were sampled at a higher-than-normal rate and weighted to reflect their proportion of the population.

In addition to sampling error, the practical difficulties of conducting any public opinion survey may introduce other sources of error. Variations in the working of the order of questions, for instance, can lead to different results [13].

Even this otherwise excellent model lacks an essential item of information. There is no statement of the response rate! As we noted earlier, there are times when a moderate proportion of nonrespondents can poison the survey well.

But Don't Turn Your Back

Development of Survey Methods to Assess Survey Practices is a publication of the American Statistical Association which succinctly gives a professional's viewpoint of surveys of human populations.* The findings and discussions are relevant to all types of surveys.

Among the observations, based on a pilot study of surveys, reported in this publication, are:

1. Fifteen of 26 federal surveys did not meet their objectives. Technical flaws included high nonresponse rates, failure to compute variances [basic to determining standard errors] or computation of variances in the wrong way, inclusion of inferences in the final report that could not be substantiated by the survey results, no verification of interviewing, and no data cleaning.

2. Seven of the 10 nonfederal surveys did not meet their objectives.

3. Samples were, for the most part, poorly designed.

4. Survey response rates were difficult to collect and compare.

5. Quality control over data processing operations varied considerably.

6. One would hope that survey results bearing on controversial subjects or public policy would include enough evidence to support reported findings. This is not generally the case. Causal inferences were often made in

---

*Available from the American Statistical Association, 1429 Duke Street, Alexandria VA 22314-3402, 703-684-1221

cases where variables were correlated in such a way that causal relationships could not have been determined by a survey of the type used (14, p. 13).

Fortunately, this publication includes recommendations on how to conduct a survey as well as criticisms based on its studies of existing surveys. We encourage you to think critically about surveys and their results, using the observations above as your guideline.

SUMMARY OF CHAPTERS 9 AND 10

Common Pitfalls and Cautionary Recommendations

A great part of the statistics which the public sees and which affects public policy is based on surveys. Someone formulates questions, decides on a sample, and asks those questions of people in the sample. There is much room for misuse of statistics in this process.

Below are six major pitfalls and our suggestions to the reader who wants to avoid survey misinterpretations:

1. *Leading questions—those that consciously or unconsciously are designed to obtain desired answers.* The way to deal with this pitfall is to find and evaluate the original questions. Much good statistical survey work is being done and the major professional pollsters and many publications make a point of giving the exact wording of the questions. If the result of the survey is important to you, it is worthwhile seeking out the questions, even if some extra effort is required. For guidance, see Chapter 9.

2. *Surveys that are really requests for contributions and attempts to gain "scientific" support for one viewpoint.* These surveys usually have leading questions linked to requests for contributions. Such so-called surveys are not surveys and cannot, and should not, be given the credibility that you give to a professional survey or one in which the surveyor has tried (even if only partially successfully) to be unbiased. However, there is information in the results of such a misuse of statistics, for the nature of the responses tells you something about a particular group of respondents.

3. *Self-selected samples.* Self-selected samples tell you only about the characteristics of those who have selected themselves. In general, the results cannot be projected to any other population.

4. *Nonrespondents.* Know what the response rate is. Determine how the nonrespondents might affect the observed results. Good surveyors establish the characteristics of the nonrespondents to estimate what effect they might have. Look for such efforts in the report. In our final example below, we show how to deal with nonrespondents.

5. *Failure to take into account the effects of variability.* If the report does not give a measure or estimate of variability, then compute one as best as you can. If the sample size or other information needed to compute vari-

ability (for example, the standard deviation) is not available and the survey results are important to you, ask for them.

6. *Misinterpretation of the results, either in the report or in subsequent media reporting.* You can avoid this pitfall by reading the full body of the report and checking for logical consistency.

Are We Asking Too Much?

In a real situation, no survey will be perfect. Nor do surveys need to be perfect; they only need be good enough to serve the purpose for which they are undertaken. Imprecision and inaccuracy may be tolerable; it is only important to know "how good is good enough."

One of the most frequent sources of inadequacy in surveys is, as we have mentioned, a high rate of nonresponse. If a survey has a high proportion of nonrespondents, there is always the possibility that the results might have been significantly different if the nonrespondents had responded. Below we describe a case in which the surveyors struggled, and succeeded, in overcoming this problem, providing a useful model for other surveyors.

In a study described in *Negro Higher Education in the 1960s* (15), the survey population was all of the black high schools in states where less than 8% of blacks were attending high school with whites in 1964–65. The first mailing of questionnaires went to 1831 high schools. After this mailing the surveyors discovered that some schools had closed, others had changed their grade levels, and some on the original list were white schools. The returns from the first mailing were so low that a second mailing was made. Only 203 high schools responded to the first two mailings and:

> Given the low initial rate of response, it was beyond our financial means and time limitations to secure returns from the two thirds or better of all the schools necessary to assure a reasonably representative sample.
>
> Instead, we attempted to obtain a 100 per cent response from a sample of fifty of the schools on the list. [These responses] would provide an index of the representativeness of the larger sample when compared to . . . the larger group of returns. . . .
>
> After extensive follow-up mailings, plus numerous phone calls, we managed to account for all fifty high schools. For forty of them we obtained useful questionnaire returns, five were White high schools, and five were no longer high schools [15, pp. 204–205].

These results were then systematically compared with the results from the larger sample with its low response rate, and the final data were adjusted to account for the information gained from the sample of the nonrespondents. All of this occurred under conditions of limited financial means and time. If the will exists to make sure that the survey is not deficient, the means and the time can be found.

11
The Law of Parsimony: Ockham's Razor

Out of the clutter find simplicity.
Out of discord make harmony.
Out of difficulty find opportunity.
                                    —Albert Einstein

You go on the tennis court to play tennis, not to see if the lines are straight.
                                    —Robert Frost

INTRODUCTION

William of Ockham (1280–1349) was an English scholastic philosopher. In an era when highly convoluted arguments were used to prove how many angels could dance on the head of a pin, he proposed his principle of parsimony (now also known as "Ockham's razor"): "What can be done with fewer [assumptions] is done in vain with more." Unfortunately, modern statistics is often in need of a shave.

Translating Ockham's principle of parsimony for the world of statistical analysis, it says: the simplest procedures that can be used to solve a problem are the preferred ones. Those who deliberately complicate solutions by using complex procedures when simpler ones will work just as well are guilty of a misuse of statistics. Such misuses may bolster the ego (See how smart I am, everyone!). They may obscure an uncertain analysis (I don't want anyone to see how shaky my work is). Sometimes, these misuses even lead to a job promotion for the perpetrator (See how smart I am, boss!):

> Complexity and obscurity have professional value—they are the academic equivalent of apprenticeship rules in the building trades. They exclude the outsiders, keep down the competition, preserve the image of

a privileged or priestly class. The man who makes things clear is a scab. . . .

Additionally and especially in the social sciences, much unclear writing is based on unclear or incomplete thought. It is possible with safety to be technically obscure about something you haven't thought out. It is impossible to be wholly clear on something you do not understand. Clarity thus exposes flaws in the thought. The person who undertakes to make difficult matters clear is infringing on the sovereign right of numerous economists, sociologists and political scientists to make bad writing the disguise for sloppy, imprecise or incomplete thought [1].

In an amusing satire, Edi Karni and Barbara K. Shapiro describe what can happen when statisticians disregard the principle of Ockham's razor:

[The investigation of the Committee on the Mistreatment of Raw Data— COMRAD] unmasked such nefarious schemes as employment of third-degree autoregression processes and, in what may be the report's most revolting disclosure, the brutal imposition of third-degree polynomial structures. Maximalist methods also include the use of first and even second differencing, which according to eye witnesses, often reduced the data to a totally unrecognizable state [2].

EXAMPLES

The best way to illustrate the principles of parsimony is by example. The following examples make all too clear the unshaven state of some modern statistical work. Unfortunately, much of this statistical overkill is in part fueled by the ease of using computers, which facilitates the use of sophisticated methods.

A Lack of Sufficient Statistical Discrimination

Naive use of statistics (and attendant complexity) is illustrated by the report of the Ad Hoc Committee to Implement the 1972 Resolution on Fair Employment Practices in Employment of Women of the American Anthropological Association (3). The committee's purpose was to find out whether some academic anthropology departments discriminated against women in their hiring practices. The known facts were: (a) for the years 1973–77, the proportion of all doctoral degrees in anthropology awarded to women; (b) the proportion of all hirings that were women, school by school, in departments of anthropology; and (c) the proportion of faculty members in each school who were women.

A nonstatistician (or the owner of a suitable razor) would have simply compared the proportion of women in all hirings, school by school, with the proportion of doctoral degrees awarded to women overall. The investigator would then know which schools hired a smaller proportion of women than the national average. Then you can search for the reasons why these schools have a lower proportion to determine if discrimination has taken place. A neat, close shave by Ockham's razor.

Instead, the committee devised its own measure. Someone decided to cross-multiply the two percentages: the proportion of women in the faculty and the proportion of newly hired faculty members who were women. This is comparing oranges with oranges, for the numbers must overlap. The resulting statistical tables were most imposing and may have misled people. We hope that some people were able to look behind the complex presentations and draw reasonable conclusions. The naive statistics and unnecessary complexity made the task more difficult (4).

In the Digs

No field is immune. Even archeologists have disregarded Ockham's principle. Writing of this, David H. Thomas, of the Anthropology Department of the American Museum of Natural History, describes how archeologists have used—and overused—complex statistical methodology:

> Too much of the methodological "new archaeology" seems involved with adapting show gimmicks, which function as symbols to attract adherents and amass power . . . simulation studies, computer methods of typology, and the numerous "models" that blight the archaeological literature of late—random walks, black boxes, Markov chains, matrix analysis, and so forth. . . . Quantitative techniques can (and do) provide valuable tools when properly used. Specialized quantitative methods become bandwagons . . . when they become ends in themselves, playthings that serve as roadblocks to real understanding [5, p. 235].

As an example of how the law of parsimony was violated, Thomas describes Joel Gunn's analysis of C. Melvin Aikens' original findings, published five years earlier, discussing an excavation at Hogup Cave in Utah (6). Gunn stated that the purpose of his reanalysis of Aikens' data was to determine "to what extent environment causes cultural change," and Gunn is quoted by Thomas as saying, "It was considered important to keep the substantive aspects simple and to concentrate on methodology." This is our first warning signal!

Thomas describes how Gunn "transformed (Aikens' data) by $\log_{10}$ to reduce skewing effects of very high artifact counts, standardized to a mean of zero and a standard deviation of 1, then similarity coefficients were computed and factored." Thomas goes on to note that the statistical relationship that emerged "was patently obvious to all who had read Aikens's analysis."

Why was it processed and massaged to such great lengths when the statistical conclusions were not altered?

Next, Gunn carried out factor analyses and multiple regressions on the data. Thomas says, "The present analysis is so far removed from archaeological reality that the numbers take on a life of their own. And the conclusions offer us nothing new" (5, p. 240). Indeed, Thomas was truly amazed at one of the major conclusions, which stated that "habitat causes a cultural change of 50%." What on earth did this mean? The archeological specifics that Aikens had considered had been transformed (by Gunn), through statistical maneuvering, into archeological numbers largely without meaning. Gunn's analytic overkill illustrates a too-frequent academic misuse: Concentrate on the methodology, the substance be damned.

Analyses of Nonexistent Data

Death Rates

Statistical analysis of death rates is another field where Ockham's razor could be used. Here, it is often the lack of good original data which seduces statisticians into overusing the data that are available.

If death rates can be calculated, then it is possible to calculate a life table and measure the average number of years the newborn can be expected to live. But in many parts of the world it is impossible to get the basic data necessary for calculating death rates. If the basic data—the number of people in each age category and the number of deaths that occur in each category—are unknown, we cannot calculate death rates. What is to be done?

The demographer W. Brass devised an analytical method to determine death rates which involved asking women about the numbers of their dead and living children (7–9). The proportion of dead children (of all born), when combined with knowledge of the age of the mother and the ages of the children, gives the death rate. The life expectancy at birth can then be estimated, using a series of complicated procedures. This description is an oversimplification, but it does describe the general methodological idea.

There are many assumptions that must be made to support these perhaps too complex analytical computations. Michel Garenne gives a list of eight such assumptions (10). We feel that the basic problem with this methodology is that the respondents must give *correct* information about their ages, the number of children born to them, and the number who died, according to age at death. We will not go into the problems with the other assumptions here.

Garenne proposed a closer shave in his article. After carrying through all the necessary Brass method computations, he concludes:

> In this example from Tropical Africa, it seems that at this level of knowledge, data on proportion of children dead do not permit the computation of reliable estimates of infant and child mortality. . . . Inaccuracy of data on age as well as underestimation of children deaths will give no more than a rough idea of the level of mortality. . . . *Why not stay with the proportions of children dead? At least by themselves, they give an idea of the level of mortality* [italics ours]. . . . As long as there are no accurate data in Tropical Africa, one cannot expect to have more than "ideas" of levels and differentials in mortality [10, p. 129].

We are not casting doubt on the Brass methodology in general. This methodology need not be shaved if a country has adequate data for the application of the Brass methodology and lacks the necessary data for computing death rates. But, for countries with less than adequate data, a close shave may be the best that can be achieved.

Answers Without Data

Wassily Leontief, winner of a Nobel Prize in Economics, was troubled by the tendency of economists to build models without using actual data. He analyzed the articles which appeared in 10 years of issues of the journal *American Economic Review* and found that over half were:

> . . . mathematical models without any data. Year after year economic theorists continue to produce scores of mathematical models . . . without being able to advance, in any perceptible way, a systematic understanding of the structure and the operations of a real economic system [11].

We agree—models that are not tested with data constitute misuses. Leontief goes on to comment:

> Thirty-six percent [of research papers studied] also contain attempts at empirical implementation of these intricate theoretical constructs. Such attempts involve routine application of elaborate methods of indirect statistical inference applied to a small number of . . . indices. . . . Only two of the 44 researchers saw fit to engage in the grubby task of ascertaining by direct observation how business actually arrives at [its assessments of government behavior and consequent effects] [11].

The "lack of substance and precision in much current economic literature" bothered Richard Staley, another observer:

> I once described this phenomenon to some graduate students as a procedure of piling estimate on top of conjecture, declaring the whole to be an axiom based on the author's reputation and then using this "base" to launch still further estimates and pseudo-precise conjectures. . . . Personally, I tend to opt for the philosophy that less may be better under these circumstances [12, p. 1204].

Hybrids

Sometimes overcomplexity is mixed with oversimplification to produce a hybrid misuse. Gerald A. Bodner was involved in a court case involving the difference in salary between a group of male faculty members and a group of female faculty members (13). The female faculty members claimed that they were receiving lower pay even though their qualifications were just as good. Were the salary differentials due to intentional sex discrimination? The plaintiffs' study produced regression equations which included as many as 98 independent variables. Do enough data exist to support that many variables? Any formula that attempts to cover that many variables indicates the need for a closer shave with Ockham's razor.

On the other hand, the court approved an analysis which included oversimplification. Information that should have been included in the salary analysis was omitted. Bodner writes:

> Pre-hire years of teaching experience and pre-hire publications, both clearly relevant and important factors, were left out because, as the statistician for the plaintiff suggests, with the court's acquiescence, they are "adequately accounted for by the variables of age, degrees, and years between degrees." The accuracy with which data on age and degrees measure pre-hire publications and years of teaching experience may be a good indication of the accuracy of the entire statistical case [13].

The principle of Ockham's razor does not imply that relevant facts can be omitted. Proxy or substitute information instead of actual data can constitute an oversimplification, especially when the actual data can be obtained—as in the case with prehire publications and teaching experience. We aim for the simplest approach, but not at the cost of disregarding relevant information.

Less Is More

Wesley A. Fisher tried to find out whether marriage between persons of different ethnic groups in the Soviet Union was influenced by ethnic consciousness, or by social and demographic factors (14). He analyzed 14 different ethnic groups in the former Soviet Union and defined eight characteristics with which he hoped to explain the amount of intermarriage between persons of different ethnic origins. He analyzed the data using multiple linear regression to obtain an equation for predicting the amount of intermarriage (see Chapter 7). He judged his results to be definitive since, by using all eight characteristics as independent variables in the regression equation, he accounted for 74% of the variance.*

---

*This concept is discussed in detail in Chapter 7. You may recall that the amount of variance explained by the independent ("explanatory" or "predictor") variables in a regression equation is the coefficient of determination, R^2.

Reviewing Fisher's results, Brian D. Silver of Michigan State University showed that five of the eight independent variables included in the regression equation added nothing to the final explanation (15).

One serious difficulty with Fisher's original analysis is the use of only 14 cases for which data were available (that's how many ethnic groups he had), to support an equation with eight independent variables. It is well known that, as the number of cases and the number of independent variables approach each other, the reliability of prediction by the independent variables approaches zero. To give an idea of how close Fisher is to the practical limit with eight independent variables and 14 cases, note that if the numbers of cases and independent variables are the same, then the proportion of variance explained will always be 100%.

Silver's recalculations from the original data agreed with Fisher's results. He then rejected the five independent variables which were not significant in explaining the variance and computed the resulting regression equation based on only three independent variables. This equation, shaved down from eight independent variables to three, explained about 79% of the variance, compared to Fisher's 74%.[*]

Silver analyzed Fisher's results and concluded that, because two of the three independent variables which indisputably contributed to the explanation of intermarriage were religion and native language, ethnic consciousness, *not* social and demographic factors, was most important. This directly contradicts Fisher's findings. Indeed, people in the former Soviet Union tended to marry coreligionists; religion alone explains 55% of the variance in intermarriage. A close shave produced a better result than the much more complex analysis. Are you surprised?

More Can Be Too Much

In the previous example, explaining 70% to 80% of the variance was considered a significant indication of the ability of independent variables to predict. "Would you believe 99.9969% explained" is the title of a provocative paper by two Dow Chemical Company researchers who used a complex computer program to get a regression equation which they hoped would lead to a solution of a quality problem (16).

---

[*](This footnote is for readers with a modest familiarity with multiple regression.) You may be surprised to see that a reduction in the number of independent variables increases the explained variance. Usually, adding another independent variable increases the proportion of explained variance, but the increase can be insignificantly small. However, when the number of cases is small and close to the number of independent variables, the "adjusted" coefficient of determination must be used, and it is possible for a reduction in the number of independent variables to increase the adjusted coefficient of determination. Less can be more!

Their odyssey in Ockham's never-never land began with their involvement in a production quality problem. A customer found that a chemical product had an unpredictable service life and questioned the quality of the batches. Representatives of the company's relevant operations (production, engineering, research, quality control) assembled and agreed to collect all the available data "and submit it to the computer for analysis." Additional data were also collected. Sixteen independent variables were defined; the dependent variable was service life of the product. Data were collected for 22 batches. Additional "independent variables" were created by performing transformations on the original 16: the reciprocal of each original variable, the square, and the reciprocal of the square. This gave a total of 64 independent variables.

The computer program for analysis of these data is a "stepwise" multiple regression program. To explain the variance, this program moves in "steps," adding independent variables to the linear multiple regression equation one at a time.* At each step, the program adds to the equation that one variable from the remaining set of independent variables which explains the greatest amount of the (as yet) unexplained variance. The program also drops previously entered variables from the equation when the inclusion of a new variable makes them less important. In the Dow case, this complex analysis resulted in the following finding: about 75% to 80% of the variance in service life was explained by independent variables (16).

But these same two researchers were suspicious of the whole complex process. So they drew numbers out of a hat at random to get values for all 16 independent variables for the 22 batches. The transformations of these data made for a total of 64 independent variables. When they submitted these random data to the same computer program, they found that 99.9969% of the variance in service life was explained by these randomly generated values! This is significantly higher than the 75% to 80% that aroused their suspicions in the first place, and threw the whole process into doubt, for, "Of course, with results like that out of a hat, it was no longer reasonable to make extensive plant revisions" (16, p. 45).

Shaving off their beards, the authors offered some sensible guidance:

> Pressures to use the latest mathematical tools in conjunction with complex computer programs are often great. Response to this pressure is evident in the number of training programs available to people engaged in quality control work.
>
> Such people, however, are in a very real dilemma: They have need to learn and to use these modern techniques, while at the same time, they have the responsibility to see that these techniques are not abused.
>
> This responsibility is often extremely difficult to meet in a practical way—that is, without resort to statistical jargon [16, p. 46].

---

*We discussed the use of regression to explain variance in Chapter 7.

Or Less Is Certainly Good Enough

In the 1960s, during the Vietnam war, one of us worked on the design of an extensive weather forecasting system for military applications. The system went into the field as a complete laboratory contained in a standard trailer truck. It had sensing devices and gave forecasts based on complex analysis of a wide range of observed atmospheric phenomena. Over 10 years later, one of our students reported on his experiences in Vietnam:

> I would like to interject [into a report on statistical forecasting methods] a short story which demonstrates what is, possibly, the true strength of the naive model.* When I said I did not call the naive model by its name, it is not that I just happened on the process by accident and didn't even think that such a process existed. My term for the naive model was persistance [sic] and it was an important part of my life for a year. The fact is, we are assailed by the persistant [sic] model every time we listen to a weather FORECAST. Let me briefly relate the story.
>
> From January 1971 to January 1972, while serving in the United States Air Force, I was assigned to the 1st Weather Group, Tan Son Nhut Vietnam. 1st Weather Group had responsibility for producing the weather forecast for Southeast Asia and providing that information to all commanders in support of war operations. As chief of weather equipment maintenance I used to sit on all the forecast briefings. Two briefings were given. The first briefing was the persistance [sic] briefing, it rained yesterday over DaNang, it will therefore rain today over DaNang. The second part of the briefing was forecaster analysis. This took into account all of the information which was obtained from what, I assure you, was the most advanced weather equipment known to science. [This was the equipment described earlier.] *The weather forecast improved upon persistance [sic] by 0.02 percentage points* [our italics].

Weather forecasting has come a long way since the early 1970s. But at that time, for the purpose intended in that place, would doing less have been more?

Doing Little in a Big Way

Some years ago, one of us attended a lecture by a market research consultant on the use of a fairly sophisticated methodology, conjoint analysis, for analyzing customer product preferences. After showing the original data, the consultant presented the methodology and the results of the analysis. At the

---

*The *naive model* means that you forecast the next period value with the current period value. For example, if a restaurant using the naive model wants to forecast the number of customers to be served for the next day, it uses the number served the previous day as the forecast value.

end, a member of the audience said, "But wasn't it possible to draw the same conclusions from a simple analysis of the raw data?" Conspiratorially, the consultant replied, "Well, I have found that my clients like a bit of mystery with their answers." So much for the idea that it is always the analyst who wants to use the most "sophisticated" methods.

As stated in an article reporting on its possible use, this sophisticated method (conjoint analysis) is "presumed" to have "superiority ... over simpler, less expensive techniques" (17). The authors of the article say it all when summing up their comparison of the complex conjoint analysis relative to two more direct, simple, and basic methods:

> Careful evaluations over time of the comparative quality of the market-place predictions produced from these two basic approaches can provide the marketer with the basic information from which to determine accurately which measurement techniques work best under various conditions. This knowledge, in turn, will enable the marketer to avoid using an expensive, complex and time-consuming methodological technique when a simple, inexpensive, easy-to-understand approach might provide as good, or perhaps better, data [17].

Robert Lewis of Columbia University called the tendency of researchers and analysts to emphasize method rather than the subject, "deification of methodology." For some people, the drive to use the most sophisticated method is overwhelming. As we show below, such a drive is often coupled with a lack of knowledge of the subject. As Dr. Douglas A. Samuelson says,

> The combination of attempted methodological sophistication and ignorance of the subject is almost certainly not accidental. There are two primary causes, in my experience: (1) people trying to obscure their ignorance by shifting attention to their methods and (2) people who are primarily interested in promoting their pet methods and don't care about the subject. Perhaps there's a variation on a well-known old joke here: Those who can, do. Those who can't, develop general methodologies which allegedly would enable others to do, and then run around urging others to use their methods [18].

Robots at Work

In a paper published in a prestigious peer-reviewed journal *Technometrics*, three academic authors without apparent connection to engineering departments addressed the problem of "establishing the *causal relationships* [our emphasis] between speed, weight, and distance and the performance measures of repeatability and accuracy" (19).

Some idea of the authors' woeful lack of technical expertise may be had from the following statement appearing in the Introduction: "Due to gravity, humans experience some fatigue when the arm is stretched farther away from the body, and one would expect the same to be true for industrial robots." But fatigue in metal comes from repeated bending, not from being extended while carrying a heavy load.

In laymen's terms, the authors want to determine if the robot can repeatedly place its arm at the desired point and with what precision (scatter of placements with respect to some central point) with different distances, weights, and speeds. It is certainly reasonable to want to test a commercial robot to determine such performance measures based on tests. Testing of this kind is often a part of the process of placing new equipment in use.

However, the authors do not just want to test or calibrate the robot; they want to determine the robot's "causal relationships." They propose to do this by using structural equations, a sophisticated social science modeling process, which they generate from data using a LISREL®* causal model. This is a useful tool where there is no direct causal basis for analysis, as in "economics, sociology, psychology, political science, marketing, epidemiology, and education" (20). It involves analyses of covariance matrices and in this particular case led to a "General Structural Model" containing 16 nodes and about 50 connecting line segments. The authors' "managerial implications" include the following:

> The objective should therefore be to choose a robot that meets the minimum desired attributes specified for the planned operation, plus a given amount of buffer to allow for efficient operation and possible future requirements in the production system. . . . it would seem wise to operate robots at the minimum speed necessary to perform the task, rather than at the limit of the robots [sic] speed capabilities. . . . Management should use manufacturer-specified information cautiously. . . .

No engineering or manufacturer manager needs a complex model to deduce this advice. Worse yet, no engineer or physicist needs a structural equation model with its uncertainties and complexities to determine the repeatability and accuracy of a robot under varying external conditions. The response of the arm is determined by well-known laws of dynamics and the nature of the mechanical structure and propulsion mechanism of the robot. These laws *are* the causal equations; the engineer's problem is to apply them in a given case. If a fairly straightforward closed analytical model (formulas based on the laws of dynamics) cannot be developed because of nonlinearity or other analytical issues, then a simulation model can be developed.

Monkeying Around with Business

An executive of a large American corporation supplied us with an internal report on the company's unrewarding experience with a team of consultants

---

*LISREL® is a registered trademark of Scientific Software.

who obviously needed Ockham's razor (21). The authors found many flaws, which they called "statistical fallacies." We excerpt from their report concerning the "most serious single defect":

> The fundamental model is . . . a regression, in which return on investment [the dependent variable] is "explained" by about 70 "independent variables," consisting of quantities obtained from the questionnaire, their ratios, and several expressions (some very complex) built from these ratios. [It] claims to explain 78% of the "variance" in return on investment from one business to another.

When results concerning the company's capital structure as it applied to the several operating divisions were received, the corporate staff spent several months

> . . . together with the appropriate business analysts, attempting to understand the structure of the model and to trace the path from our input data to the final results. We found that [several terms in the model] played an overwhelming role, both in the general structure of the model and in its application to our businesses. The difficulty was that the [results] did not agree well with the qualitative judgments of [the] analysts.

What the company's analysts discovered was that data of many kinds were input to the complex computer analysis system and were used to produce a regression equation which produced a value for the ratio of income to investment. But one of the company's analysts observed that:

> . . . as a matter of high school algebra . . . the ratio of income to investment can be calculated exactly from the [formula, using a few numbers available through routine accounting functions]. It need not be estimated from any model. The formula does not require empirical validation, and incorporates no empirical experience or wisdom. It is an accounting identity.
>
> The official . . . model "explains" [this accounting ratio] in terms of . . . a large number of "explanatory variables." The model does not include [the exact formula] but it includes many expressions which are *more* complicated. If that particular "interaction" term had been included, the model would have accounted for 100% of the variation in profitability from one business to another, and not merely 78%.

If we have a shortage, it is a shortage of Ockham razors, not computing capability.

12
Thinking: Lack of Forethought, Lack of Afterthought

THINK.
> —IBM motto for many decades

The great tragedy of science—the slaying of a beautiful hypothesis by an ugly fact.
> —Thomas Huxley

INTRODUCTION

Some years ago, we gave a talk on uses and misuses of statistics to an audience of faculty and students at New York University. This was a time of great concern about "comparable worth," the concept of bringing women's pay into agreement with men's on the principle of equal pay for equal work. One of us (Jaffe) discussed a news story which reported recent statistics showing that the pay of men and women had been equalized in a segment of the health care industry. The headline and story stated that this was a signal instance of bringing women's average wage rates up to the level of men's average rates.

Jaffe obtained and analyzed the data for this case and found that, in fact, the women's average wage rate had not been raised to the level of the men's, but that the men's average rate had dropped below its previous value, and the women's average rate had increased above its previous value. Essentially, the equalization of rates was the result of a decline in the men's rate along with a rise in women's average rates—and they met in the middle!

The audience reaction was unexpected. They were not concerned with the economic implications. What they wanted to know was, "How did you know to ask the question?" They wanted to know the nature of the thinking process that led to discovering the conflict between the assumed result (women's average wage rates have been raised to the same as men's) and the actual situation. To us, it was clear that this was an issue of "thinking."

These queries led us to focus on the need for examining the thinking process as it applies to the use and misuse of statistics. This is a specialized topic, which we will illustrate with a series of examples. We are not concerned with thinking as a neurological process, or "creativity" as it is usually conceived. We are concerned with thinking in the sense defined by David Kerridge, "By 'statistical thinking' we mean the framework of ideas that makes statistical techniques relevant, understandable and useful" (1).

This kind of thinking plays an important role in the evaluation of research results and the fact-finding that plays a large part in the debates over public policy and the management and administration of business enterprises. And at a more mundane daily level, it plays a key role in decision making by individuals.

We are not going to try to define our concept of thinking for this chapter more precisely. It will be implicit in the examples that follow. However, we will categorize the kinds of thinking that may have created the misuses we observed.

THE RED VELVET FLAG

Some things just jump out from the printed page as patently absurd. The observant journalist, editor, or reader can spot these absurdities, which are often obvious even to persons who have no specific knowledge of the subject. Here is one such example. While the consequences of the reporter's failure to think were not fatal, we can imagine the reactions of the state's business leaders.

"High Costs Are Blamed for the Loss of a Mill," was the headline of an article in *The New York Times* that described why a venerable 100-year-old textile company (American Velvet) was abandoning Connecticut to move south to Virginia (2). In giving the reasons for this move, the *Times* reporter noted that one component in the decision was the high cost of employment: "Personnel costs—wages, worker's compensation and unemployment insurance— are *20 times* [our emphasis] higher in Connecticut than in Virginia."

One of the authors read this and a red velvet flag went up. If the personnel costs in Virginia were indeed 20 times less than those of Connecticut, the giant sucking sound of Connecticut jobs heading south would have been heard around the world.

When isolated by us, this conclusion is clearly absurd. But absurd as it is, it originated with a reporter and apparently passed through the *Times'* editorial review processes without a question. When we contacted the reporter, she insisted that her figures were correct. After all, she had taken them from a cost analysis supplied by the company president.

American Velvet gave us a copy of the cost analysis given to the *Times* reporter. To be fair to the reporter, as the analysis was presented it could easily mislead someone without a knowledge of business cost analysis. (This is discussed in Chapter 4, Know the Subject.) We do not expect individuals in the editorial process to know business accounting practices; most journalists do not have MBA degrees. Our concern is elsewhere; it is *the failure to recognize an absurd result*. This, then, is one of the major topics in this chapter on "Thinking."

In this case of not thinking, the reporter had simply misread the report. She mistook the word "TOTAL" in the report for the total of wages, workmen's compensation, and unemployment insurance, when it was, in fact, a *subtotal* of unemployment insurance costs. This one cost—unemployment insurance—*was* 20 times higher in Connecticut, but wages were about 1.2 times higher and worker's compensation was about six times higher. The total of the values for these three components ("personnel costs"), is 1.31 times higher than in Virginia. This is bad enough, but it is not 20 times higher.

Of course, there were other costs such as real estate, energy, pollution controls, etc. These costs, whether higher or lower, undoubtedly played a role in the decision, but the members of the editorial staff should have recognized the absurdity of a ratio of "20 times." This takes *thinking*. Thirteen days later, the *Times* printed a correction, ending the possibility of a mass migration of Connecticut companies to Virginia (3).

THE JANUS EFFECT

The Roman god Janus was blessed with two heads, facing in opposite directions. This gave him an unusual perspective, as described in the quote below:

> Clearly [the] double-headed fetish at the gateway of the negro villages of Surinam bears a close resemblance to . . . the double-headed images of Janus which, grasping a stick in one hand and a key in the other, stood sentinel at Roman gates and doorways; and we can hardly doubt that in both cases the heads facing two ways are to be similarly explained as expressive of the vigilance of the guardian god, who kept his eye on spiritual foes behind and before, and stood ready to bludgeon them on the spot [4, pp. 166–167].

Look Both Ways Before You Cross the Street

We are indebted to Robyn Dawes, University Professor at Carnegie-Mellon University, for the following examples of a common and often serious category of misuse involving Not Thinking. We give some examples of this particularly dangerous misuse involving estimates of probability based on statistics, or capable of rectification with statistics.

Our first example will illustrate the nature of the problem.

Use Your Speller

The admissions committee of an Ivy League university wants to increase its diversity by admitting entrants who are "neat small-town kids" (but still meet its admissions standards). A young woman applying to study engineering fits the desired pattern. She comes from a poor rural area, and has excellent test scores in mathematics. Unfortunately, she spells "engineering" wrong. "Dyslexia," says a member of the admissions committee (rather than "carelessness" or "poor spelling skills"), and the committee puts her on the waiting list.

Alas, this member of the committee looked "only in one direction." Observing that the applicant had misspelled a word, he used the high probability he derived from his experience (accumulated informal data) that dyslectics often misspell words to conclude that there was a high probability that the applicant was dyslectic. He failed to think about the opposite "direction" and ask the question: "What is the probability of misspelling by nondyslectics?" As Dawes says, "There are many more of us who cannot spell well who are *not* dyslectic than who *are*" (5).

Surgeons Beware

Sometimes this failure to think in "both directions" can have serious consequences. About 20 years ago, a newspaper article reported that a surgeon was taking a "pioneering approach to the treatment of breast cancer" by removing "high-risk" breasts before cancer developed. At the time, about 50% of tested women fell into the high-risk group, and over 93% of breast cancers occurred in the women who are in the high-risk group. Based on this reasoning, in two years the surgeon removed the noncancerous breasts of 90 women whose mammograms placed them in the high-risk group. We will show that the physician made the Janus error, and based his decisions on reasoning "in the wrong direction." In order to do this, we must reconstruct the research data that he used.

Whether a woman has a high breast risk for breast cancer was based, in this case, on a radiologist's research which showed by mammography that certain "patterns" seen in the breast served as indicators of breast cancer risk. Debate about whether to attempt to determine if a woman is in the high-risk group and how to proceed continues at this time (1997). From the description of the work of the researcher from whose work the surgeon determined his probabilities, Professor Dawes reconstructed Table 12.1 for 1000 "typical" women.

Table 12.1 Reconstructed Complete Table of Counts
for 1000 Women Based on Surveys

Breast risk factor	Cancer		
	No	Yes	Total
High	499	71	570
Low	424	6	430
Total	923	77	1000

In Table 12.2, we separate out the column for those women who developed cancer and compute the percentages. You can see from Table 12.2 that at that time, of the 77 women who developed breast cancer 71 (or 93%) were in the high-risk group.

But this unidirectional view merely tells us that the great majority of the women developing breast cancer are in the high-risk group. *It doesn't tell us what proportion of the women in the high-risk group develops breast cancer!*

To get that answer, we have to look in the opposite direction. In Table 12.3, we separate out the *row* for the high-risk women, and compute the proportion of women in the high-risk group who develop breast cancer.

You can see that although 93% of the cancers occur in high-risk women, *the proportion of the high risk women who will develop cancer is only 71 out of 570 (or 12.5%)*! Does a risk of 12.5% justify removal of noncancerous breasts? This is a decision for each individual and her physician who must consider if such a radical operation is justified to avoid a 12.5% risk?

Unfortunately, it is not unusual for physicians to misuse statistics this way. Two legal scholars found that 90% of the physicians they surveyed made this error when dealing with diagnostic tests (6). Patients beware!

Lawyers Beware

This kind of misuse is called *the prosecutor's fallacy* when a witness, juror, attorney, or judge commits it. In their book on interpreting evidence, Robertson and Vigneaux report on the testimony of a psychologist which concluded that

Table 12.2 Partial Table of Counts and Percentages
for Cancer Sufferers; Group Percentages

Breast risk factor	Yes (number)	Yes (%)
High	71	93%
Low	6	7%
Total	77	100%

Table 12.3 Partial Table of Counts and Percentages
for High-Risk Group Percentages

Breast Risk Factor	No	Yes	Total
High (number)	499	71	570
(%)	88.5%	12.5%	100.0%

a child had been sexually abused partly on the basis of the alleged victim's report of a dream of a kind frequently reported by sexually abused people (7, pp. 27–28). No valid conclusion can be drawn without a knowledge of how many people who have *not* been sexually abused have such dreams.

Avoid this error by keeping in mind the two-headed god, the Roman Janus.

WILD AND WOOLLY STATEMENTS

You can say, write, or print anything, but that doesn't make it true. Many off-the-wall statements can readily be tested, and far too many beautiful hypotheses are slain by ugly facts. Here is one beautiful hypothesis deserving of execution by facts:

The Thought Is Admirable, the Statistics Are Not

This is the statistical statement appearing on a bumper sticker: "U.S. SPENDS MORE IN 5 HOURS ON DEFENSE THAN IN 5 YEARS ON HEALTH CARE" (8).

A quick look at the *Statistical Abstract* will provide the facts that slay this hypothesis. It is easy to determine annual defense expenditures, since they come from the government and appear in the federal budget and subsequently in the *Statistical Abstract*. The estimated total national defense and veterans' outlays in 1994 are reported to be $319 billion. We add in veterans' outlays to attribute the largest possible amount to spending on "defense" (9, Table 539).

But who is the "U.S." that spends so little on health care? We look only at federal health expenditures to give the author of this bumper sticker the benefit of the doubt. Total *federal government* spending for health services and supplies totaled $290 billion in 1994 (9, Table 159).

Thus we see that federal health expenditures alone are roughly the same as the expenditures on defense. If you take the "U.S." to be the country as a whole (public, private, federal, state, and local government), then the total U.S. health expenditures is $402 billion, 26% *more* than the defense expenditures. The statement on the bumper sticker is a statistical lie, and easily proven to be so. Do the people who put this sticker on their bumpers and the people who read it think about what it says?

Totting Up the Togas

Even scientists are not exempt from the wild and woolly virus. An item discussing the Ebola virus and the Plague of Athens in the June 14, 1996, issue of *SCIENCE* makes the statement that "up to 300,000 Athenians—one in every three—were felled during a Spartan siege by a mystery disease." As Richard Ellis points out in his subsequent letter to the editor, if this were so, then the number of Athenians must have been about three times 300,000 or 900,000 (10). This is an unbelievable number. The best estimates we can get for the population components of Athens at the time of the Spartan siege follow.

The authors of *The Cambridge Ancient History* conclude that, allowing for women and children, total population of Athenian free citizens at the time was about 150,000 to 170,000, to which can be added 35,000 to 40,000 resident aliens. They estimate that the slaves numbered 80,000 to 100,000 but are unlikely to have exceeded 120,000 (11, p. 200). If we take the upper limits of all these estimates we find that the highest reasonable estimate of the total population is 330,000.

Had 300,000 Athenians died of the plague, Athens would have vanished from history.

ASSUMPTION, ASSUMPTION, WHO'S GOT THE ASSUMPTION?

In the comparable worth example with which we started this chapter (see the Introduction), the breakdown in thinking had a frequently observed cause. The headline writer, the reporter, and our audience all made the *assumption*—without explicitly stating it—that if men's and women's wages were equalized, the women's wages must have been increased while the men's stayed constant. As we saw, this assumption was invalid as the men's wages were reduced and the women's increased less than the original difference between the two wage rates. Here's another case of a misplaced assumption.

Where's the Job?

The number of jobs lost is always a key indicator of economic security for the country. It takes on a special importance during an election period, when the party in power likes to show steady economic growth. There was a great deal of concern about job losses as the country entered the 1996 Presidential election period.

Anecdotal information supported the feeling that job losses were rising, but in its August 1996 release of the estimated involuntary job losses from 1993 to 1995, the U.S. Labor Department's Bureau of Labor Statistics announced a decline to 8.4 million, 7% less than the estimate for the preceding period from 1991 to 1993. A month later the same Bureau of Labor Statistics announced its work on a revision of the estimate which would lead to a result showing that at best, job losses did not decline and quite likely rose. In October 1996,

> The U.S. Department of Labor "acknowledging a flaw in a key statistics in the debate over economic insecurity . . . said today that it would revise upward its tabulation of the number of layoffs in the mid-1990s after an academic researcher found a problem with the Government's figures [12]."

How did this happen? Henry Farber, a Princeton University labor economist, found that the proportion of nonrespondents in the current 50,000-household survey was 9%, an increase of 7.2% over the prior survey. The Bureau of Labor Statistics stated that it assumed that the percentage of nonrespondents had not changed from survey to survey when making its calculations: "The Bureau had considered whether it needed to adjust for nonrespondents but decided not to do so in the mistaken belief that the percentage of nonrespondents had been the same in both surveys" (12).

This mistaken belief—we would call it an assumption—could have easily been checked. Yet no one in the Bureau of Labor Statistics *thought* to check it.

LET'S BE PC: PERCENTAGEWISE CORRECT

The percentage is a simple, useful statistical tool for summarizing, comparing, and, occasionally, relating data. We make no wild and woolly estimate as to the percentage of Americans who are percentagewise-disadvantaged. However, the number of examples that float to the surface and regularly reach us suggests that there is a widespread failure to think when creating and judging percentages.

Let's Scotch the Budget for Police

How can you beat this for quality of life? A Scottish statistician who lives near Ayrshire sent us this example of nonthinking.

In 1993, the *Ayrshire Leader* (Scotland) newspaper, reported: "Local police are particularly pleased with decreases in the number of robberies, break-ins and car-related crimes. In fact robberies in the area are down by a staggering 100%" (13).

A decline of 100% brings robberies down to zero. There is no way that a shire as large as Ayrshire (about 370,000 inhabitants in 1991) could be zero, even if the reporting period were as short as one week. The probability of having zero robberies in any given reasonable reporting period is vanishingly small.* This example reflects the failure to think about the absurdity of a 100% reduction in an area this large. Watch out for failures of this type associated with percentages.

Now Here Is a Real Epidemic

Edward MacNeal, in an article in *ETC: A Review of General Semantics,* reports the following headline appearing in the Waterbury (Connecticut) *Republican-American:* "92% of Young Men Have AIDS Virus" (14).

Once again, we have a statement which is so patently absurd that both the headline writer and the copy editors should have recognized its percentagewise incorrectness (PI). Is it possible that PI is a contagious statistical disease, and close proximity to the infected leads to clusters of severely infected individuals? As it turns out, the first sentence of the accompanying news story reveals the error: "One of every 92 young American men—those ages 27 to 39—may be battling the AIDS virus, according to the most precise estimates yet of the epidemic's toll" (14).

We can see how inadequate thinking led the headline writer to an incorrect percentage of 92% when the true percentage was 1.1% (1/92). All is not lost, for a subsequent correction appeared in the newspaper: "A study reported in Friday's *Republican-American* showed that approximately one American Male in 92 was infected with the AIDS virus. A headline in the Litchfield edition misstated the number of people infected."

Like wolves, misuses usually run in packs. As MacNeal observes, this correction itself has two errors. It unjustifiably extends the population under discussion from American men ages 27 to 39 to *all* American males—by implication, those of all ages. Also, the original headline did not misstate the *number* of people, it misstated the *percentage*. The contagious nature of the percentagewise-incorrect virus might explain why so many people in the same organization can have difficulty thinking clearly about percentages. Did this

---

*Assuming a Poisson distribution with mean robbery rate of 5.3, the probability of zero occurrences is 0.005.

misuse pass through the editorial review process? And if it did, why didn't someone see the absurdity in the statement?

Running With the Percentagewise-Incorrect

Maybe a Cat Would Be Safer

Even scientists who certainly should know better can be PI. In 1997, a team of researchers reported that they had found that the dog had only one ancestor, the wolf, and that domestication of the wolf must have occurred much earlier than believed. Both supporters and critics are dogged by doubts about the estimate of the date of domestication, since it was obtained by the use of a "mitochondrial clock" which uses cellular information from DNA to estimate an approximate date (15).

The nominal value of the date is 135,000 years ago, but many scientists caution observers that this form of data is not reliable: "The date is very dubious—it's 135,000 years plus or minus about 300%," says [Stephen] O'Brien [geneticist and chief at the Laboratory of Genomic Diversity at the National Cancer Institute]. Did anyone think through what he said? Three hundred percent of 135,000 years is 405,000 years. This give a range of the estimate of time from −270,000 to +540,000 years. The upper limit is not unreasonable, but are we expected to wait another 270,000 years for the wolf to be domesticated? If so, we should get them out of our homes as quickly as possible.

OLD MISUSES NEVER DIE, THEY DON'T EVEN FADE AWAY

It is a consequence of the definition of the arithmetic mean (see Chapter 7) that the mean will lie somewhere between the lowest and highest values. In the unrealistic and meaningless case that all values which make up the mean are the same, all values will be equal to the average. In an unlikely and impractical case, it is possible for only one of many values to be above or below the average. By the very definition of the average, it is impossible for all values to be above average in any case.

Unfortunately, that does not stop people and organizations whose livelihood depends on being above average from claiming to be so. We will show you cases in which this claim is made. Don't jump to the conclusion that this is deliberate, and not a failure to think. We will show you that even people who try to correct this error think incorrectly about averages.

Are All the Children Above Average?

The national radio show "A Prairie Home Companion" (Garrison Keillor) has publicized the mythical small town of Lake Wobegon, Wisconsin, in which "all the children are above average." Keillor meant this statement as a joke; he obviously knows enough about the average to know that it is impossible for every value to be above average, and the audience laughter that greets this statement shows that a large part of the audience also knows this is dead wrong. After all, we use the average to give us a "typical value," and it hardly makes sense (thinking!) for every value to be above the "typical" value. The mythical Lake Wobegone school system must have administered a test of some desirable attribute to its children and computed the arithmetic mean as an average score.

If the Lake Wobegone children are compared to a *national* average, they might well all be above that average. This would be something for Wobegone's school system to brag about. Unless they are trying to justify additional funding to improve performance, most school districts want to announce that their students are performing above "grade level." This appears to show how well they are doing their jobs. In ways that vary from region to region and from time to time, "grade level" is defined as the average of a *nationwide* test of students in a given grade. Clearly, all the students in the nation could not be above average. But *all* the students in a given school could certainly be above the national average.

The Chancellor of the New York City school system, Dr. Rudy Crew, set a goal of having every third-grader read at or above grade level. This is a reasonable goal if the comparison is against the national standard, or even the national average. Unfortunately for the children and the chancellor, the third-grade students in New York did not achieve this goal in 1997. In a letter to the editor of the *New York Times*, a reader accuses Chancellor Crew of a "seemingly shaky grasp of statistics":

> On the test, "grade level" is defined as the average score of a nationwide sample of third graders, so 100 percent of the city's third graders would have to beat the national *average*, an impossible goal [14].

We don't know if Chancellor Crew has a problem with statistics, but we do know that the letter writer's grasp of statistics is shaky, without any qualification. Did anyone think to challenge this statement?

The letter writer's statement is incorrect on its own merits. It would certainly be possible for the students to exceed the national average, by the very definition of average.

However, there is a question whether there *is* a national average. Today, in 1997 as in the past, there is no national test for reading and mathematics. A proposal for such tests is "now [mid-1997] being developed by the Education Department, [and] is under attack from the left, right and middle" (17).

In 1987, John Jacob Cannell, MD, President of Friends of Education, picked up on this issue (18). His contribution was to make it clear that (a) the standardized tests being used "do not use a current group average for comparison, but rather use a 'norm group' from the past for comparison, so it is statistically possible for 100 percent of current students tested to be above average"; (b) the methods of testing the students vary from district to district and state to state; (c) the administered tests may be modified versions of the tests; (d) not all districts and states make their results known publicly; and (e) reporting methods differ. These are the kinds of issues that should enter into thinking about this subject and illustrate why knowledge of the subject is so important to clear thinking.

Fortunately, Dr. Cannell received a great deal of media attention, and this helped to open up a window on the process. But this was a window that some of the companies producing the tests tried to keep closed. In the words of Dr. Peter Behuniak, Director of Student Assessment of the Connecticut Department of Education, they "stonewalled him because they didn't want to admit the existence of problems in the process." However, post-Cannell, they have tried to improve the process by shortening the cycle time of testing to establish a norm and the release of tests. Nevertheless, norms may be two to four years old when comparisons of current results are made. As long as the members of the media and public who rejoice in "above-average" reports do not raise questions about the testing and norming process, they will draw invalid conclusions from the results. And the students will be the losers.

It is too bad that Dr. Cannell also engages in some muddled thinking about statistics. He believes that "Standard principles of mathematics make it difficult for more than half of any group to be above the group average." As we discuss above, any nonzero part of the group can be above or below the average. In addition, at times he seems to be talking about the median. By the definition of the median, it is *always* true that the measurements for a group are split equally on either side of the median. At other times, in making his arguments, he appears to assume that all test measurements are distributed symmetrically about the average.

Improvement in statistical education and thinking is slow in coming!

CUTTING A GOOD FIGURE: WHO'S COUNTING?

A New Epidemic?

Do you want to make a scientific estimate of a proportion? As we point out in Chapter 10, you can do this with a sample if the sample is randomly chosen from the target population. It is a statistical no-no to go from a finding based on a handful of anecdotes, or a convenience sample (the subjects that you find easy to get), or from a sample with a high proportion of nonrespondents, or any nonrandom sample, to a generalization about the whole population. But this is a no-no routinely ignored by many people, especially those who want to make a newsworthy splash. In an article on self-mutilation in America, Jennifer Egan, a novelist and writer, quotes Dr. Armando Favazza of the University of Missouri as saying that he estimates that there are 750 per 100,000 *Americans*, or "close to 2 million" who are self-mutilators, and he "suggests" that the actual figure may be even higher (19).

The thinking reader should immediately note that almost all of the cases reported in the article are anecdotes about adolescent females. What credence can we give this figure of 2 million? On what is this large estimate based? The only large-scale survey mentioned in the article consists of a population of 240 American females, probably ranging in age from 14 to their late 20s. The red flag is that the figure of "2 million" is based on the "estimate" of 750 per 100,000 as extrapolated to the *entire* American population of 250 million, when, in fact, the only study mentioned—which may or may not be statistically sound—is based on one segment of the population, females from 14 to 29 years of age. There may be another study which is more scientifically based, but when we spoke directly to Dr. Favazza, he could not reference such a study.

For the nonthinking reader, this new threat to adolescents might seem ready to eclipse the teen-age diseases of the month, anorexia and bulimia, but before it does, we count on physicians and researchers to do a better job of counting.

THERE IS NOTHING EITHER GOOD OR BAD, BUT THINKING MAKES IT SO

"Our customers seem to think that our products don't compete in the high-quality marketplace," "Spousal abuse is on the increase," "The Internet is becoming a household necessity," "Meteorites could be a threat to planet Earth," "Ethical behavior is on a decline." These are expression of problems, the vague yet often important concepts that trigger the creative process. The people who are concerned with resolving such problems usually define statistical artifacts that give substance to the expression of the problem.

For example, an executive of a national fast-food company may decide that potential customers are not coming to their restaurants because of "low product quality." What statistical data can the company use to measure the perceived quality of their products? They may choose to use a statistical artifact as simple as the proportion of potential customers who respond "lower" to a question asking whether the respondent thinks the products are of higher, equal, or poorer quality than the competition. Or, they may set up a set of questions about specific attributes of their products (such as serving temperature, seasoning, flavor, presentation, and so forth). Social science provides a formal structure for relating the measurements and the concepts.

However, "commonsense" thinking serves many people in many disciplines well in moving from a *problem* to *questions* that can be answered by statistical results. Thinking is more important than any formal procedure.

Is This the Good News?

In the elections and public debate of the 1990s, no issue got as much lip service as the welfare of American children. And for good reason. Of particular concern was the teenage mother, considered the indicator of all that was wrong with society, and in particular, the "broken" welfare system inherited from the New Deal. Having promised to do something about this problem, members of the Clinton administration found reason to brag. The administration put the word out that the teenage birthrate was now in decline: "On Feb. 28, 1996, Donna E. Shalala, the able secretary of health and human services told the Senate Finance Committee [that] 'teenage pregnancy rates have gone down'" (20).

This decline in teenage birth rates was quite modest (down from a high of 60.7 when the administration came into office in 1992, to 56.9 in 1995). Unfortunately, "What is the trend in teenage birth rates" is not the right question to ask if you are concerned with the problem of the welfare of children.

It seems to us that the teen birth rate is secondary in importance to the *illegitimacy* rate in the welfare of children. Remembering that a 19-year-old is defined as a "teenager," we might assume that there are at least some present members of the Congress and other upstanding members of society who are the children of teenage mothers. It is the children without a father and the related emotional and economic support who represent an issue of child welfare.

You can see from Figure 12-1 that the teenage illegitimacy ratio (the proportion of teenage births that are illegitimate) has been on the rise since 1970. If indeed there is a reduction in this ratio, it will auger well for the welfare of children.

SUMMARY

Having reached this point, you will know what we mean by thinking. It is a kind of thinking logically or quizzically that affects many areas of our lives.

Teenage Illegitimacy Ratio vs.Time

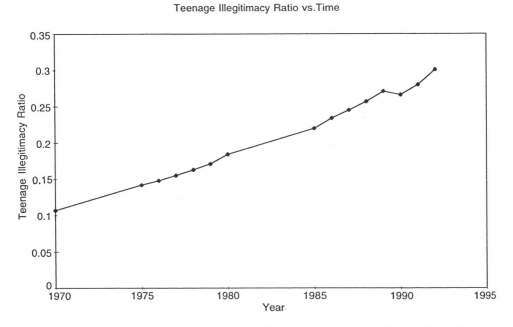

Figure 12.1 The ratio of illegitimate births to teenagers to the total number of teenage births (9, Table 97).

But we are only interested in those areas in which numbers are used or misused. We do not have to emphasize the importance of numbers to you, the reader. If you did not believe that numbers and measurement were important to the conduct of life and the world's business, you would not be reading this book.

We are not the only people who have concern for "thinking" about statistics as a special type process. An article published in *Science* on teaching reasoning discusses statistical reasoning as one of the major topical areas (20). The authors' findings agree with our own anecdotal observations: that statistical reasoning (thinking) can be taught: "Taken together, the results of our studies suggest that the effects of higher education on the rule underlying reasoning may be very marked. . . . It is indeed possible to improve inferential rules through training" (19, pp. 630–631).

The authors also conclude that their work supports the concept of learning through experience, a process "in which intelligence evolves in the context of everyday problem solving." It is in this latter sense that we have written this book: to contribute to the reader's development of statistical thinking skills. Thus, we sum up our examples using the following incomplete but hopefully useful set of guidelines (some of which have been described in earlier chapters):

Implicit Assumptions

Take nothing for granted. Don't assume that the questions asked in a poll reflect the purpose of the survey; find out what they are. Don't assume that the experimenters "must have" controlled the conditions of the experiment. Don't assume that a complex methodology is correct because of its sophistication.

Know the Subject or Get Someone Who Does

Know the subject matter, learn it fast, or get a trustworthy expert. To identify the unknown, you must know the known. But don't be afraid to challenge experts on the basis of your logical reasoning. Sometimes a knowledge of the subject matter can blind the expert to the novel or unexpected. How much subject matter knowledge is enough? As an answer we offer Abraham Lincoln's response when asked "How long should a man's legs be?" He replied, "Long enough to reach the ground."

All Statistics Are Artifacts

Clear definitions are rarely given in casual reports. Peer-reviewed research is no guarantee of clarity—we have seen many papers that have incomplete, missing, or inconsistent definitions. The range of reporting on indicators of importance to the public policy debates varies from the superb to the abysmal. Even the people generating the results may have trouble getting their definitional act together.

Refuse to Be Snowed

Jargon and complex methodology have their place. But true professional jargon is merely a shorthand way of speaking. Distrust any jargon that cannot be translated into plain English. Sophisticated methods can bring unique insights, but they can also be used to cover inadequate data and thinking. Good analysts can explain their methods in simple, direct terms. Distrust anyone who can't make clear how they have treated the data. We have seen all too many instances where the mere use (correct or incorrect) of numerical data is enough to get people to accept a result without questioning. Don't be one of those people!

Make Your Own Rough Estimates

You don't need a supercomputer to check statements such as that on the bumper sticker described earlier in this chapter (U.S. spends more in 5 hours on defense than in 5 years on health care). In many cases of nonthinking, high school arithmetic is enough to pop the balloon.

Get Your Own Copy of the Statistical Abstract of the U.S.

If you care about issues which concern public policy, or business enterprise, order a copy each year. Then you can check a lot of the "facts" which have managed to escape any legitimate thinking process.

Take Nothing for Granted

13
Ectoplastistics

ec'to-plasm, n. [ecto-, and -plasm.] . . . 2. in spiritualism, the vaporous, luminous substance supposed to emanate from the medium's body during a trance.

—Webster's New Twentieth Century Dictionary

I'm not sure how many yuppies used to be dweebs or nerds. (They are not old enough to have been weenies or wonks.) The last statistic I heard was 38 percent—although as I understand it, that figure was arrived at by factoring in a wimp quotient, which I find completely irrelevant.

—Calvin Trilling

INTRODUCTION

Spiders and humans have an ability in common: spiders spin silken threads from their bodies to catch unwary insects; humans spin thoughts—both myths and statistics—with which to trap the unwary hearer. Even though some of these human emanations are "vaporous" and "luminous," they are often accepted by the unwary as "the truth."

Is the real difference between *Homo sapiens sapiens* (to distinguish ourselves from long-gone ancestors) and its predecessors the ability to produce numbers? If so, it is both a great advantage and disadvantage, for numerical emanations are sometimes emitted without serious regard for their truthfulness. Thus are created **ectoplastistics** [n., ecto-, -pla, and -s(ta)tistics], which can be used to get legislation passed, raise armies, create antagonism among peoples, extol or slander, win political position, change economic activity, enhance careers, get customers, and gather media attention.

As you will see in the following examples, no human activity is immune from the influence of ectoplastistics. You must be constantly on your guard, and where the professionals do not do your ghostbusting, you will have to do it yourself.

GHOSTBUSTING

Ectoplastistics are ubiquitous. You can't bust them all. If the subject is important to you, then have a go at bursting the bubble. We recommend the following steps.

Identify the Apparition

A likely clue is spurious precision, the statement of a remarkably precise number where none could conceivably exist. Here's an example. In discussing his work on deaths caused by governments, R. J. Rummel states that his partial total "of this century's megamurders total 126,424,000 men, women, and children. . . ." (1)

This implies that he knows the number of murdered to within 1000. The number is not 126,425,000, or 126,423,000, but exactly 126,424,000. To some, this number rounded out to the final three zeros may seem to have been rounded to a reasonable value. But the way this number is given is actually a statement of a precision of less than a thousandth of 1%. It implies that the total number of deaths from megamurders can be known to a thousand deaths. Despite all the records and research, we cannot even know the exact numbers of the dead in the Holocaust, or in the Gulag camps of the former Soviet Union, to such a precision.

The lack of substance of the ectoplastistic specter is not always so evident. Often, a clue to an ectoplastistic number will be the absence of any reference to an estimating method or a source. Sometimes you have to have some knowledge of the subject to be suspicious. Living in the New York metropolitan area, we have some familiarity with the regional statistics. Rhoda Howard, in *Human Rights and the Search for Community*, gives us an example of ectoplastistics put forward by critics of life in modern, capitalistic Western states:

> These [critics] . . . hold what might be called "The Central Park Thesis," the logic of which is as follows. . . . The United States is characterized by very high crime rates. They are highest in New York, a center of acquisitive materialism and greed [2].

Dr. Howard goes on to criticize the Central Park Thesis. Our concern is with the belief of proponents of the Central Park Thesis that New York has the highest crime rates in the U.S., a belief held by many Americans as well as foreigners. If you suspect that a statement of this type is wrong, you can readily check it in the *Statistical Abstract*. Many ectoplastistic numbers can be verified in this excellent source or other appropriate references.

So how high *is* the crime rate in New York City compared to other regions? According to the *Statistical Abstract's* ranking of the largest cities of the U.S., based on the total crime index, New York is 55th (3, Table 313).

Exorcise the Phantom

Common sense may be enough to blow away the ectoplastistic demon, as we saw in the case of the estimate of the world's megamurders. Finding a credible source for the true number is also effective in clearing out such emanation, as in the example of New York's crime rate. Knowledge of the subject may be all that is needed in some cases (which you will see later in this chapter).

However, if these methods fail to clear out the ectoplastistic phantasms, you can go direct to the source. This takes persistence, but if the ghost is important to you, it is worth the effort. Be prepared to find that the source of the ectoplastistic statement ignores your phone calls, letters, faxes, and e-mails. In general we do not forgive them, for they usually know what they are doing. You may be able to contact the originator, or second-best persons who know something about how the ectoplastistics was generated. We show this process in some of the examples which follow.

WHEN NUMBERS ARE VICTIMS OF VIOLENCE

The [Stamford, CT] *Advocate's* house seems to be haunted with ectoplastistics, especially in regard to abuse of women. When discussing the O.J. Simpson case in January 1997, author and publisher Janus Adams states, "Horrifying statistics tell us that 75 percent of American women will endure gender-related abuse, assault—or worse" (4). Common sense alone is enough to puncture this numerical balloon.

No Passing Grade

In an article concerning women's health published in the [Stamford, CT] *Advocate,* there were 10 questions (5). The reader was asked to choose the correct answer to the questions, one of which was "How many Connecticut women are victims of violence each year?" The possible answers were:

a. 100,000–150,000
b. 150,000–200,000
c. 200,000–250,000
d. 250,000–300,000

The "correct" answer was stated to be (d), 250,000 to 300,000. Connecticut is a small state with about 3 million residents; in 1995, about 1.7 million were women. Could 15% to 18% of Connecticut women be victims of violence each year?

According to the *Statistical Abstract* (3, Table 312), in 1994, crimes of violence against both women *and* men in Connecticut totaled about 14,000. A nongovernmental organization, the Connecticut Coalition Against Domestic Violence (CCADV), stated that in 1995, police reports of violence totaled 17,256, of which 13,709 were women (6). In addition, the CCADV report included estimates of violence against women through hot lines, shelter services, nonshelter services, and criminal-court-based services. Adding these additional instances, the total is about 75,000 (and there may be duplicate counts). Bad enough! But hardly 300,000.

Where did a number of 250,000 to 300,000 come from? We tried to contact the public information officer of Connecticut's governmental Commission on Women which supplied the story to the newspaper. Their public information officer did not reply, but the Commission sent us a publication with these same figures with no sources.

Continuing the search (as we said, ghostbusters must be persistent), we reached another state organization, the Domestic Training Project, where the contact person said that she *thought* the number had come from the former head of CCADV, who left to join an organization concerned with women's' affairs in another state. We kept digging and found a spokesperson for the Connecticut Department of Public Health who said that her department used the police statistics for their numbers of women victims of violence. We asked both parties if they knew whether 300,000 was an annual estimate; none were sure.

Conclusion? Forget the 250,000–300,000 figure. This frightening number launched on an unsuspecting public is just another ectoplastistic emanation, ephemeral and insubstantial. A definitive published report containing reliable data and valid conclusions would lay this ghost to rest.

Divorced from Reality

No-fault divorce was a major issue in the 1970s. Under the no-fault rules, marriages can be terminated without either spouse having to prove that they have been in some way wronged by the other spouse. There has been considerable political attack on the practice, and some states are considering doing away with no-fault divorce.

One of the important arguments of those pushing for elimination of this type of divorce is the claim by Lenore Weitzman, a sociologist and author of a 1985 book on divorce, that in the first year after a no-fault divorce the man's standard of living rises 42% and the woman's drops 73% (7).

Other social scientists have consistently found different values—a drop for the women between 13% and 25%, and a rise for the men between 11% and 13%. For example, in *The New York Times,* author Nancy Rubin quotes a 1989 joint study by the Connecticut Women's and Education Fund, the Women's Research Institute of Hartford College for Women, and the Permanent Commission on the Status of Women which found that mean per capita income of Connecticut women dropped 16% following a divorce while that of men rose by 10%. According to Ms. Rubin, a more recent study appearing in the *American Sociological Review* reports that women's income dropped by 27% while that of divorced men rose by 10%. These figures are part of an ongoing refutation of Weitzman's statistics (8).

When would-be ghostbusters (other social scientists in the field) requested access to Weitzman's data, they were politely stonewalled. In 1993 (five years after publication), she made the data available to other researchers. One of them, Richard Peterson, recoded and analyzed her data and found a 27% female drop and 10% male increase in the standard of living (9).

When Weitzman released her data, she warned of serious errors and problems. She stated that among these problems was the fact that she and another researcher could not replicate her original results. Peterson's work is a superb example of good statistical reasoning and analysis. His conclusion was that the "most likely explanation is that errors in her analysis of the data were responsible for producing the results" (9, p. 534). In a reply to Weitzman's comments on this paper (10), Peterson summarizes:

> Weitzman's inaccurate estimates were widely circulated, and the conclusions drawn from them have seriously distorted policy discussions about no-fault divorce. Weitzman . . . contends that her analysis of the law in action, not her erroneous statistics, were [*sic*] responsible for the attention [her book] received. . . . However, it is misleading for her "to put in perspective the 73/42 percent datum" by arguing, "It is one statistic in a 500-page book." . . . These figures are featured prominently on the book jacket, as well as in the introduction and the concluding chapter. Most reviews . . . cited the figures as a major finding, *as did Weitzman herself* [in Congressional testimony]. . . .

Yet when she realized that there were problems with her figures, she did not publish or disseminate a warning that she could not replicate them (11, p. 540).

Stay Awake and Guard Against Ephemeral Emanations

No day passes without some new attention-getting threats to humanity. Real or imagined, they often are publicized by the emission of industrial-strength ectoplastistics. Are you sleep-deprived? This was the topic of a lead article in *The New York Times*. If you believe the ectoplastistics on this subject, there is a high probability that you are a tired reader. Here's what the author of the article says: "Almost everyone I know complains about sleep, and the refrain is usually 'Not enough [sleep].' It's a subjective estimate, but accurate as far as it goes" (12).

The author is right about one thing; this is a subjective estimate. But of what? It evaluates only the author's set of friends, who must be a most somnolent daytime lot, since almost every one of them complained about not getting enough sleep. From here it is all downhill, as the author continues:

> The problem of sleep curtailment in late-20th-century Western society "is so big," one prominent sleep researcher told me, "that people just can't digest it. If you were to take people off the street, the vast majority would be sleep-deprived. There is a sense among many students of the subject that sleep deprivation is reaching crisis proportions. It is a problem not only for serious insomniacs, who total perhaps 17 percent of American's adult population, but also for the populace at large [12].

What proportion is the "vast majority"? Eighty percent, ninety percent? A real estimate of this proportion would call for a random sample of the American adult population, administered with the same care as other credible social science surveys. No such source is given, and the statement seems to reflect no basis for this estimate.

As for the "serious insomniacs" who *perhaps* number 17%, what are we to make of this estimate? This is a variation of the spurious precision problem. Seventeen percent; not 18, not 16; but still "perhaps." If this is the way these "researchers" talk, can we trust their work? Clearly, these numbers cannot be taken seriously. Perhaps these researchers need a good night's rest.

The Great Lobbyist Infestation: The Birth of an Ectoplastistic

As you have seen, some ectoplastistics seem to arise because they satisfy the media's interest in the spectacular and frightening and because the originator has a great deal to gain from putting an ectoplastistic number forward. In a situation where a reasonable estimate does not exist, it is not uncommon for a journalist, columnist, politician, or other leader to misspeak deliberately or through ignorance.

In the following well-documented case, we can see a typical ectoplastistic-generating process at work. In 1993, President Clinton said that 80,000 lobbyists "are lining the corridors of Washington as never before" (14). We can trace the development of this spooky figure from its birth to full adulthood: The figure was born when a reporter for the *Wall Street Journal*, Jeffrey Birnbaum, contacted Mr. James Thurber [a professor of government at American University] for an article on efforts to revise lobbying laws. Specifically, he wanted to know how many lobbyists there were in Washington. Mr. Thurber recalls: "I said, 'Jeff I don't know. 60,000,70,000, I'm sure 80 is a reasonable figure." Later, Mr. Thurber said, "I got worried about it because I did it off the top of my head" (13).

Uncomfortable that his estimate "was becoming truth," Mr. Thurber said, he recently did a more systematic study. He took a sampling from major directories of advocacy groups to determine the number of lobbying groups as well as the number of staff members associated with each. "I've come up with a little higher figure," he said; "91,000 lobbyists and people associated with lobbying activities in and around Washington" (13).

As we discuss in Chapter 3, all statistics are artifacts and have meaning only in reference to a specific definition. The lobbyists lining the corridors in Washington are not cooks, secretaries, and others "associated with lobbying activities." But altering the definition of lobbyist to include them gives a much higher estimate, and makes the original number look a lot less embarrassing—but not more correct.

Others concerned at the time with the number of lobbyists were no less ectoplastistic than Thurber. Reported estimates from unknown sources run between 6000 and 20,000. The executive director of the American League of Lobbyists said that her high estimate would be 10,000. A journalist at the *Washington Monthly* said that 40,000 "was probably more accurate." But then, he added, "the reality is usually somewhere in between." That statement should be in the Ectoplastistic Hall of Fame.

Ectoplastistics all, because there is not an agreed-upon and consistent definition of a "lobbyist," and no systematic survey based on that definition. Only one fact turns up in this whole discussion: 6085 people were registered as lobbyists under the law at the Office of Records and Registration of the House of Representatives (13).

Just How Much Does Addiction Cost?

Around 1970, it was commonly believed that New York City's heroin addicts committed about half of all property crimes. Many prominent individuals and organizations supported this view which stated that the amount of property theft by heroin addicts was between $2 billion and $5 billion per year. Is this an ectoplastistic emanation or an estimate based on fact? According to Max Singer, who followed the ghostbusting policy of getting to the source, the estimators made assumptions and worked as follows:

> There are 100,000 addicts with an average habit of $30.00 per day. This means addicts must have some $1.1 billion a year to pay for their heroin (100,000 × 365 × $30.00). Because the addict must sell the property he steals to a fence for only about a quarter of its value, or less, addicts must steal some $4 billion or $5 billion a year to pay for their heroin [14, p. 3].

How does this reasoning stand up? If we accept the assumption that the total number of addicts is 100,000, how many addicts steal? Some of the addicts can pay for all or part of their habit without stealing. In addition, there are data which support the premise that the street addicts who steal spend about one-quarter of their "careers" as addicts in jail. This reduces the estimate of the number of addicts at any one time to about 75,000, according to Singer.

The figure of $30.00 a day is based on the amount of heroin consumed and its street price. Is it reasonable? An unknown number of heroin addicts are in the trade and get their heroin at a wholesale cost significantly less than $30.00 per day. Addicts steal cash as well as property, which means that they can sometimes get full value for their thefts without going through a fence. These are unknowns, so other methods of valuation must be used.

Singer asked, "What happens if you approach the question from the other side?" Mr. Singer has obviously heard of the Janus effect. "How much property is stolen by addicts or anyone else?" Addict theft must be less than total theft. What is the value of property stolen in New York City in any year?" (14, p. 4).

Singer could not get any supported estimate of total theft. He therefore made his own estimate based on fragments of information available, which we do not repeat, but which you can judge by reference to his article. His conclusion:

> If we credit addicts with all of the shoplifting, all of the theft from homes, and all of the theft from persons, total property stolen by addicts in a year in New York City amounts to some 330 million dollars. You can throw in all the "fudge factors" you want, and all the other miscellaneous crimes that addicts commit, but no matter what you do, it is difficult to find a basis for estimating that addicts steal over half a billion dollars a year, and a quarter billion looks like a better estimate, although perhaps on the high side. After all, there must be some thieves who are not addicts [14, pp. 5–6].

Thus, the $2 billion to $5 billion value is an enhanced ectoplastistic emanation, which vaporizes rapidly on examination. Singer gives good advice on how to examine ectoplastistic emanations:

> The main point of this article may well be to illustrate how far one can go in bounding [ed. note: putting limits around] a problem by taking numbers seriously, seeing what they imply, checking various implications against each other and against general knowledge [14, p. 6].

We agree with Singer's approach, which supports the major lesson of this chapter: how both ordinary people and experts can avoid being led astray by ectoplastistic visions.

Are the American People Becoming More Litigious?

"The courts in the United States are overburdened. Why? Because Americans litigate too much. They run to the courts at the slightest provocation, real or imaginary." This is a cry that is frequently heard in the United States where Tom Paxton's cynical and satirical song *One Million Lawyers* found a wide and sympathetic audience. And how often have you heard the quote from Shakespeare's *Henry VI*: "The first thing we do, let's kill all the lawyers."

To prove the contention that the United States is overburdened with litigation, complainers quote statistics. Are the statistics real or imaginary? According to Marc Galantner, Professor of Law and South Asian Studies at the University of Wisconsin-Madison and president of the Law and Society Association, the statistical evidence usually given is:

1. The growth in filings in federal courts;
2. The growth in size of the legal profession;
3. Accounts of monster cases (such as the AT&T and IBM antitrust cases) and the vast amounts of resources consumed in such litigation;
4. Atrocity stories—that is, citation of cases that seem grotesque, petty or extravagant: A half-million dollar suit is filed by a woman against community officials because they forbid her to breast-feed her child at the community pool; a child sues his parents for "mal-parenting"; a disappointed suitor brings suit for being stood up on a date; rejected mistresses sue their former paramours; sports fans sue officials and management; and
5. War stories—that is, accounts of personal experience by business and other managers about how litigation impinges on their institutions, ties their hands, impairs efficiency, runs up costs, and so forth [15, p. 12].

Even if this evidence indicates that we have a great deal of litigation, how do we know that it is too much? Do the growth and expenditure statistics (1, 2, and 3 above) and anecdotes (4 and 5 above) tell us that the cause of overburdened courts is that Americans litigate too much?

What can we say about usage of the law courts in America? There is statistical reason to think that the rate of litigation was much higher in the "good old days" of Colonial America:

In Accomack County, Virginia, in 1639 the litigation rate of 240 per thousand was more than four times that in any contemporary American county for which we have data. In a seven year period, 20% of the adult population appeared in court five or more times as parties or witnesses. In Salem County, Massachusetts, about 11% of the adult males were involved in court conflicts during the year 1683. . . . Most men living there had some involvement with the court system and many of them appeared repeatedly [15, p. 41].

Closer to our time, Galantner describes an analysis of "federal district court activity from 1900 to 1980 [which] shows a dramatic reduction in the duration of civil cases from about three and a half years at the beginning of the century to 1.16 years in 1980," and another which reports that "The number of cases terminated per judge has been steady since World War II and remains considerably lower than in the inter-war period" [15, p. 37].

For another viewpoint, consider the growth in the number of filings. Not all of the cases that are filed come before judges. Many filed cases are dropped or settled without the intervention of a judge and never reach the courtroom. The number of filings certainly is greater than the number of cases heard in the courts.

Galantner could find no consistent pattern in prior analyses and statistics to indicate a widespread large increase in the use of law courts. Unfortunately, the data he could find were limited, but some ideas can be derived from them. For example, a study of the St. Louis Circuit Court revealed that 31 cases per thousand of population were filed in the decade 1820–1829. By the decade 1890–1899, the rate had fallen to seven per thousand of population, but it rose to 17 in the period 1970–1977. In other localities for which records exist, the rate can be shown to have decreased over the last century, to have risen, or to simply have wandered up and down without any trend (15, pp. 38–41). Is the United States an exceptionally litigious country? Galantner found (16, p. 55):

> ... the United States rate of per capita use of the regular civil courts in 1975 was just below 44 per thousand. This is in the same range as England, Ontario, Australia, Denmark, New Zealand, somewhat higher than Germany or Sweden, and far higher than Japan, Spain, or Italy. ... Given the serious problems of comparison, it would be foolhardy to draw any strong conclusions about the relative contentiousness or litigiousness of populations from these data. ...
>
> The United States has many more lawyers than any other country— more than twice as many per capita as its closest rival. In contrast the number of judges is relatively small. The ratio of lawyers to judges in the United States is one of [the] highest anywhere; the private sector of the law industry is very large relative to the public institutional sector. (Perhaps this has some connection with the feeling of extreme overload expressed by many American judges.)

The real issue, according to Galantner, may be the changing nature of the U.S. society over the decades and the changing nature of the work of the courts. With the passage of laws concerning environment, health, safety, sexual harassment, race crimes, and welfare, it is inevitable that cases involving these laws would come to court. Such cases may arouse more emotional response on the part of judicial representatives of the government than the large volume in colonial times of cases between individuals, since the current cases often involve the government itself.

Ectoplastistics in Court: How Many Workers Will Get Asbestos-Related Diseases?

Since about the time of World War I, it has been known that prolonged exposure to asbestos fibers can cause cancer and related diseases. But, as Paul Brodeur pointed out (16), many key organizations ignored warnings of these effects, including the federal government, Johns-Manville (the major manufacturer of asbestos products), and firms that installed asbestos products. Apparently, thousands of workers who installed the asbestos products in buildings, factories, schools, naval vessels, and many other places did not know that inhalation of the asbestos fibers could cause asbestosis and the equally deadly mesothelioma. In the heyday of the asbestos industry, they endured these sicknesses largely in ignorance of their origin in the workplace.

Initially, a few court cases were filed by workers seeking compensation. The number of cases, as well as the amount of damages awarded by judges, by juries, or in negotiated settlements, grew. By the late 1970s, as the largest producer of asbestos products, Johns-Manville found itself on the losing side of a large number of cases. Over and over again, it was proved in court that Johns-Manville had long known that asbestos fibers were dangerous to the health of workers and deliberately chose to ignore this information. Indeed, they even attempted to suppress it, but (even in that time before computer storage of data) written documents survived in corporate files. The plaintiffs' attorneys were able to introduce them as evidence.

What was the company to do in the face of this rising tide of increasingly expensive courtroom defeats? Company officials decided that voluntary Chapter 11 bankruptcy* was a good way to protect Johns-Manville from being forced to pay out "too much" money to victims and their attorneys. But to do this, Johns-Manville had to give the court estimates of the probable future claims from pending and future lawsuits that the company would have to pay. If the total of the estimated losses was judged by the court to be too great, or if no estimate could be made, then it was very likely that Johns-Manville would not be permitted to file a Chapter 11 bankruptcy.

Therein lies the motive for Johns-Manville's ectoplastistic emanations. The company commissioned a study of the possible future incidence of asbestos-related cancer and the number of potential claims. An epidemiologist,

---

*In a Chapter 11 bankruptcy, the court prevents any collection of debts owed by the debtor until an arrangement is made to equitably distribute the debtor's available assets to the creditors.

Nancy Dreyer, was hired to do the job. She analyzed data supplied by the company and estimated a total of 230,000 additional cases of asbestos disease and 49,000 new lawsuits by the year 2000. This contrasted sharply with estimates made by Dr. Irving Selikoff, one of the world's experts on asbestos disease, who predicted 270,000 excess deaths from asbestos-related cancer alone.

The Dreyer report was honest about its uncertainties, saying, "The actual number of lawsuits might easily be as low as half or as much as twice the number our calculations suggest" (16, July 1, 1985, p. 42).

The management of Johns-Manville was less than happy with these uncertainties, which meant that this report could not be used in their request for a Chapter 11 bankruptcy. They then directed the team that made the report to "refine the estimates" and to try to lessen the range of uncertainty.

The company brought in Marc Victor, a legal decision analysis expert, to "assist in this refinement." Victor had neither experience nor training in epidemiology, but nevertheless he succeeded in

> . . . persuading Dr. Walker, a [public health] scientist, to reshape some of his assumptions concerning the future incidence of asbestos disease. . . . Walker revised his original estimate in such a way as to lower the projected number of people who might develop lung cancer as a result of exposure to asbestos . . . by discounting the risk multiplier for lung cancer that had been proposed by Dr. Selikoff . . . Walker allowed himself to be persuaded that Dr. Selikoff and his colleagues at Mount Sinai's Environmental Sciences Laboratory had vastly overdiagnosed cases of mesothelioma [16, July 1, 1985, pp. 45–46].

Here is pure ectoplastistics in its early stages of development. We have this unusual opportunity to witness the process because of the records of the extensive legal proceedings before judges and juries. Dr. Walker's own sworn testimony gives a clear description of the fabrication of ectoplastistics:

> [Question to Walker:] Were you requested at any time by Manville or its counsel to make assumptions in your estimates that would result in lower rather than higher resulting numbers?
> Walker: I was asked . . . that whenever I had the chance to choose between two equally plausible assumptions, I should choose the assumption which led to the smaller number of cases of disease [16, July 1, 1985, p. 46].

Thus, the numbers Johns-Manville presented to the court were pure ectoplastistics. More "certainty" meant taking the most favorable values from a range of values produced as the results of the most favorable assumptions.

What can and should be done in cases like this? The appropriate action is what epidemiologist Dreyer did in her original report: give the upper and lower values produced by the estimates, describe in detail how they were obtained, and report their uncertainties. Then the readers (and if need be, judge or jury) can make their own judgments what to believe and conclude. This is what finally happened in the asbestos suits; judges and juries obtained enough information about the ectoplastistic processes to make decisions. But they have that information only because of the disclosures that were forced by the plaintiffs' attorneys, a tribute to both the legal system and the attorneys' diligence. We all need to exercise similar diligence in "busting" ectoplastistic ghosts.

Prone to Ectoplastistics: Refugee Counts

Large numbers of refugees are created by wars, genocides, massacres, and the resulting civil disorder. Where these situations are accompanied by chaos, it is extremely difficult—but not necessarily impossible—to get reasonably precise counts.

Many different parties have a stake in estimates of the numbers of refugees. For those who would provide aid, the obvious need is for estimates to determine the resources to meet the basic needs of housing, food, and medical care. In addition, governments and human rights advocates may want estimates to raise public consciousness, enhance their images, raise funds, or enact legislation. International and national courts may want to estimate the scope of the crimes of perpetrators. Or, governments will want to defend themselves against charges of either creating refugees or failing to give asylum to refugees from other countries. As stated by James Bennet in *The New York Times*:

> In the fiercely political world of international relief, the estimates of refugee populations are often bent by the political aims of the estimater [*sic*], a practice that led today to charges and countercharges of cynical manipulation of human suffering [17].

At the end of November 1996, all of these parties were involved in putting forward estimates of the number of Rwandan refugees trapped in Rwanda's contiguous neighbor, Zaire. One of the critical issues at the time was whether military forces should enter Zaire to feed the refugees and protect them as they marched home to Rwanda. Through its ambassador in Washington, the Rwandan government said that only 100,000 refugees were still in Zaire. He argued that

> . . . higher projections are attempts to "justify the international intervention."

> . . . private relief agencies also had an interest in exaggerating the numbers. "Now that most of the people are going back, and they are not going to camps, it seems that most of these relief agencies will be going out of business" [17].

But the relief agencies argued that the government of Rwanda wanted the resources diverted from the refugees to itself.

What are those "higher projections"? The United Nations High Commissioner for Refugees said that the number was 700,000 refugees. The president of Refugees International said that "Substantial numbers of people are missing, in our view 600,000" (18). Where do these numbers come from?

They would appear to be in part ectoplastistic. In 1994, two years before the discussion on which we report, the United Nations concluded that there were about 1.2 million Rwandans in Zaire, "based partly on registrations at the refugee camps." In the summer of 1996, the U.N. counted "some" 700,000 in the camps south of Zaire's major lake, Lake Kivu. Militiamen kept them from counting any of the camps north of the lake.

Since it is thought that more than 500,000 Rwandans went home after this count, the U.N. believes that more than 700,000 must remain. The arithmetic process can only leave the reader wondering. We can give credence to the U.N. count of 1.2 million in 1994, but since they could not make a similar count north of Lake Kivu in 1996, how could they know that there were still 500,000 in the camps in the north? And who counted the half a million who were supposed to have returned to Rwanda "in recent days" (of Fall 1997)? The article in the *Times* includes a map that shows counts in refugee camps throughout eastern Zaire, including 150,000 refugees in camps north of the lake, where the U.N. couldn't make counts.

The quoted estimates can only be taken as ectoplastistics leavened with a modest amount of substance. If any of the concerned parties really wanted to know the number of refugees remaining in Zaire, then they should forget the ectoplastistics and make an effort to physically count the refugees. Does this mean enumerating every refugee with an army of census takers? Not at all. A count with a precision of $\pm 25\%$ might be good enough for both the relief and political purposes. Others have made credible estimates using aerial or satellite surveys. In fact, at the time of the dispute in November 1997, the United States was sending airplanes over the disputed region but stopped flights after a plane was shot at. We will probably never have a credible estimate of the number of refugees that remained in Zaire at that time.

On or Off the Books

For better or worse, the San Francisco Public Library (SFPL) has moved into a new main building. Unfortunately, the full main collection of the library could not be fitted into the new building. As one of the ways of resolving this problem, many books were "weeded"—that is, removed from the collection— and sent to dumps.

How many is "many"? According to Nicholson Baker, the SFPL "has, by a conservative estimate, sent more than two hundred thousand books to landfills." Writing in *The New Yorker* magazine, Baker quotes a librarian:

> "I'm sure that at least that many are gone . . . I would guess that maybe a quarter of them, fifty thousand, should have been thrown out. But I would guess that at least a hundred thousand shouldn't have been. And another fifty thousand that I just can't guess . . . I personally saw at least twenty or thirty thousand books when we were still in the building that were taken out in boxes and never made it back into the collection" [18, p. 51].

Baker makes a good argument that in this case, weeding is going too far, but how do these ectoplastistics help the cause? Guess, guess, guess, these numbers are all ectoplastic.

In defending the weeding project at a public meeting, a library official stated that "'a few truckloads' of books and government documents had gone to the landfills" (18, p. 60). "Newsworthily high preliminary numbers" of the old library's capacity were leaked to the press, but these numbers were based on an error in computation. As a consequence, the subsequent headline, "FOUR CRITICS OF LIBRARY MUST EAT THEIR WORDS," hurt their cause. This is an example of how dangerous it can be to put out unsupported or erroneous statistics.

Errors are hard to avoid, but to avoid ectoplastistics is easy. If the quantity of weeded books is important to the preservation of a valuable resource, it would have been quite simple to measure the magnitude of the dumping. Truckloads or boxes could have been counted, or weighted, and more reliable estimates could have been obtained.

SUMMARY

To determine whether a statistical statement is an emanation, examine the process by which it was generated. Always ask: *How do you know it?*

Two of the major issues for the reader are:

1. *Estimates.* How were they obtained? What arithmetic operations were performed? What are the corrections that were made to the input data? Who made them? Were the estimates the result of an agreement or in response to pressure?

2. *Sources.* Who made the estimates? What were the political or economic pressures on those who made the estimates? What is the established track record of those who made the estimates?

Ghostbusting is easier than you think. Most of these emanations will blow away with just a modest puff of knowledge.

14
The Body Politic:
Governments and Politicians

Every kind of government seems to be afflicted by some evil inherent in its nature.

—Alexis de Tocqueville

Practical politics consists in ignoring facts.

—Henry Adams

All governments lie.

—William S. Burroughs

INTRODUCTION

The Inca civilization, which flourished high in the Andes, predated the Spanish empire in South America. We have no evidence of writing *per se* in the archeological finds for this civilization, but we have evidence that they kept statistical records. Their numbers were recorded on *quipus,* or knotted strings. We know that their records included, at a minimum, births and deaths, amount of food produced, amount and kinds of tribute brought to the ruler, and so forth (1, pp. 331–333).

The use of statistics by governments is as old as the formation of organized societies. Thousands of clay tablets found in the "cradle of civilization" (the region of the Tigris and Euphrates rivers) recorded the accumulations and disbursements of foodstuffs and other goods. Animal flocks, laborers' efforts, movements to and from stores, and other activities we now consider commercial but which were at that time under the control of the government and the religious hierarchy, are the substance of the myriad tablets with numerical inscriptions carved on them. In the field of external relations, the records of military and acquisitive ventures, counts of prisoners taken, reports of sizes of armies, deaths, and areas lost or gained fill many stelae, panels, and graven records. This is evident in the remarkable collection of 3000-year old tablets written in "Linear B" that were discovered in the layers of destroyed materials found at Mycenae, Thebes, Plyos, and on Crete at Cnossus:

> With the translation of the tablets, it became clear that they were the accounts, or books, of the Mycenaean palace. They record, for example, inventories, assessments, deliveries, records of distribution of materials for production and of the goods produced, and particulars regarding parcels of land and flocks of animals. Additionally, the tablets preserve much incidental information regarding the political and priestly hierarchy, details of religions and cult, personal names, and so on . . . the texts are overwhelmingly administrative. . . . [they reveal] a highly stratified and compartmentalized bureaucracy, administered from the top down and meticulously monitored through a detailed accounting system [2].

Over 3000 years ago, governments collected statistics to determine how many men could be called up for military service and how many taxpayers could be separated from a part of their earnings and wealth, and to prove that their countries were becoming overcrowded and must be expanded at someone else's expense (3, p. 360 ff.).

All of this is typical of the information any government needs to maintain control, administer an empire, and allocate resources to the governed; it is the raw material of governmental bureaucracy. But it is also the raw material of the catalog of misuses of statistics by which a government can help maintain its power to control the governed while justifying its rewards and punishments, acquire wealth by extracting it from the governed, and rationalize internal repression and military and expansionary adventures.

We see these principles clearly displayed in the Inca example below. The Inca government had complete control over the governed. The Inca Empire was

> authoritarian, bureaucratic and socialistic [in a way] that has not been approached by any state at any other time or place. . . . The Inca imperial government dictated to its subjects, in detail, the locality in which they were to live, the kind of work they were to do there, and the use that was to be made of the product of their labor [1, p. xi].

While rulers may choose to use and misuse statistics as they wish, a government needs to have the straight facts for its internal use if it wants to function efficiently and effectively. We have interesting evidence of the efforts the Incas made to assure that their statistics were correct, especially at the local level, where data were created:

However small, it [the village] had at least four *quipucamayus* [the local statisticians who kept the records]. They all kept the same records and although one accountant or scribe was all that would have been necessary to keep them, the Incas preferred to have plenty in each village and for each sort of calculation, so as to avoid faults that might occur if there were few, saying that if there were a number of them, they would either all be at fault, or none of them [1, p. 331].

That the central government insisted on such strict quality control measures might be a clue that local cheating existed, since a local authority whose power is less absolute might use statistical cheating to improve his position with respect to the central government or to accumulate private wealth.

The use and misuse of statistics by governments is an ancient art. Politicians—those people who are part *of* the government and those who want to become participants—are no strangers to this ancient art. To a greater or lesser degree in all nations, governments and politicians must convince some substantial segment of the governed that certain policies are "right." For this reason, governments and politicians often turn to statistics, and use them—or misuse them—in serving their purposes.

For many of the players in this game, the stakes are high. Aside from the personal advantages to be gained (political power, high salaries, enviable perks not available to most ordinary citizens), there are individual and organizational special interests with huge financial stakes. Thus, the statistical muscle of the politician is extended through "think tanks" and study groups which can hire large numbers of analysts. Worldwide, governments command large collections of analytical resources. The few individuals who work at tagging the political misuses of statistics have often been likened to a couple of people with brooms and hand-carried trash cans following a parade of 200 elephants. The cleanup job will never be completed, as the production capacity is huge. *But we can try!*

We see at least three major schools of thought at work in misuses of statistics in the body politic. We will call them the *closet, power-drunk,* and *ectoplastistic* schools. In this final chapter, we examine some examples of each school.

The members of the closet school classify numbers as "top secret" and sequester them where the public cannot get at them. *They* will decide what numbers the general public is entitled to see. Even in the United States, such "closeting" actions occur again and again, as we know from the frequent challenges that politicians, investigative reporters, and concerned members of the public make. When one of our students sought information on public works expenditures from a city public works department, the department's representatives refused to give him permission to inspect the *public* document in which the figures had been published. The same document was available across the street in the public library!

The second school, the power-drunk, publishes and distributes statistics which it misinterprets deliberately or inadvertently in order to prove that it (the government or politician) is in the right and should remain in power or be given power.

In the ectoplastistic school, the government or politician flaunts statistics, producing them in great quantity and scattering them freely in all directions. If real numbers cannot be obtained, ectoplastistics will do:

> Capitol Hill feeds on numbers, some authoritative, some spurious to buttress arguments and gain votes. . . . many are simply manufactured and convey "an aura of spurious exactitude," in the words of Alice M. Rivlin, former director of the Congressional budget office, . . . the power of numbers in political debates has been noted by bureaucrats and lobbyists, members of Congress and Presidents, all of whom struggle to come up with the best numbers to make their cases. . . . "A person who has control of numbers, whether accurate or not, will carry the day," said Herbert Kaufman, a political scientist [4].

Since we have already discussed ectoplastistics in detail in Chapter 13, we will concentrate on cleaning out the closet.

IN AND OUT OF THE CLOSET: PRACTICES OF GOVERNMENTS

Aid and Comfort to the Enemy, or Hindrance and Discomfort to Friends?

In 1753, members of the House of Lords refused to permit a census of the British population. They argued that open knowledge of the size of Britain's population would reveal to potential enemies how small an army the British could muster (5). For this reason (and probably some others as well), the British population remained uncounted until 1801. Consequently, we have only "guesstimates" for the population of England and Wales in the middle of the 18th century, when England's industrial revolution was in its early stages (6). Because of this lack of sound statistics, it is difficult to gauge precisely the impact of the beginning of the industrial revolution on birth and death rates, on migration to and from England, and on social and economic conditions. For example, how much of the migration from England to the American colonies was due to the effects of the beginning of the industrial revolution?

We must ask: *Was this fear justified?* Would this census information really have given aid and comfort to the enemy? Or would it just have hampered the opposition? Or frustrated future historians?

Two hundred years later, in April 1944, the House of Lords repeated the same argument in a discussion pertaining to the publication of industry statistics. "We cannot relax in any way our precautions in this or other matters," said Lord Templemore. Such statistics would, once again, give aid and comfort to England's enemies (6).

This is an example of the closet school sequestering and hiding the data. There is nothing new under the sun.

How Sick Are Your Statistics?

An illustration of the problems faced by an international governmental body—the United Nations—is the way in which the World Health Organization (WHO) reported a cholera epidemic in Ethiopia in 1985. Cholera is a deadly disease if not promptly treated, and a country is supposed to report its presence to WHO. The purpose of such reporting is to alert the world community that a highly communicable disease is epidemic in some part of the world. If the country does not report cholera, WHO does not *officially* know of its existence and does not report its presence in that country. Clifford May observes that Ethiopia experienced a disease, which the Ethiopians called "acute diarrhea," and that Sudan experienced a similar disease, which Sudan called "severe gastroenteritis" (17). Western physicians said they have no doubt that these two disease reports refer to cholera:

> Officials of the World Health Organization claim to know nothing of cholera in Ethiopia and the Sudan, reflecting a provision in the organization's charter: Unless a government formally reports an outbreak, W.H.O. does not acknowledge its existence [7].

This is one of the ways in which, by the rules of their own operations, the United Nations and its member organizations can be forced to stay in the closet.

The Iron Census Curtain

No one government has a monopoly on the selective withholding of statistics, but some governments withhold more than others. The withholders may state explicitly that these statistics are vital to national security, or not state the reasons and we must impute motives to the government. There are times when secrecy in the conduct of government is justified. However, in any particular case, we must ask: Would publication really compromise national security or are the statistics withheld because they reveal a lack of good information? Or does the government withhold them so that no one will ever know whether they are good or bad?

In 1926 the former Soviet Union published quantities of population census data, but released no comparable amounts afterward. The USSR published statistics from the 1959 and 1970 population censuses, but far less than in 1926. They published even fewer from the 1979 census. Conspicuous for its absence was any information on the ages of the population.

Statistics other than those from the population censuses were equally sparse. Most published data were keyed to the Soviet republics (which were governmental units roughly equivalent to states in the United States)—data such as numbers of births, numbers of deaths, energy consumption, and so forth. Until 1974 this information was also available for oblasts (divisions of the republics roughly comparable to counties in the United States). After 1974 the USSR published no information for oblasts.

There were only 16 republics in the USSR. One of them, the Russian Soviet Federated Republic (which includes Moscow), contained over half the 262 million people in the country. Moscow, with 8.5 million people, was larger than many of those republics, such as Estonia (now an independent country), which then had only about 1.5 million inhabitants. Each republic was so diverse that little fruitful study could have been conducted on the basis of the republic alone. Information for smaller and more homogeneous areas was needed, and such information could only have been obtained from oblasts or even smaller units.

Some examples of withheld information can be seen by comparing the questions asked of the people in the 1979 population census and the data published in the slim census report. The USSR collected statistics on the ages of people but never published them; and collected data for oblasts that were never revealed. Published statistics on the numbers of children born (data needed for a comprehensive study of the birth rate) and on the educational level of the population were so sparse that almost no analyses were possible then or today. And so on and on.

We also note that any statistics which the Soviet government thought would reflect adversely upon it were rarely, if ever, released. For example, in the early 1970s, the USSR published few crime or accident statistics. When the infant death rate apparently increased, the Soviets stopped publishing such information. Actually, as Fred Grupp and Ellen Jones pointed out, the apparent increase was due to faulty statistics and not to a true increase in the death rate (8). The Kremlin should not have panicked.

The USSR did give some statistics to the U.N., which printed them in its statistical publications. The U.N. performed no analysis, and the USSR did not give enough detailed information to enable others to evaluate and interpret the totals. What the internal and external public knew about the USSR was what the government chose to let them know. Beyond that, it was all pure ectoplastistics, which included the emanations of people on both sides of the Iron Curtain.

INGENIOUS INTERPRETATIONS AND OTHER ALTERATIONS: THE POWER-DRUNK

Alter: To make different without changing into something else; castrate, spay; to become different.

>   *Alteration:* The result of altering.
>   —*Webster's Ninth New Collegiate Dictionary*

What a government doesn't suppress, it can alter. If a government cannot easily alter statistics, then it can always interpret them in such a way as to reinforce the government's position. We give a few examples in this section, limited only by space, not by the number of examples available.

Sorry About That

In February 1990, Birge Watkins, new deputy assistant secretary for food and consumer services of the U.S. Department of Agriculture, announced that the department had made a mistake in altering one of its own studies. It took a four-year battle between nutritionists and the government to get this admission. Watkins' admission leaves no doubt that governmental political appointees—the USDA Food and Nutrition director and his special assistant—deliberately "stripped summary chapters off a peer-reviewed WIC [Supplemental Food Program for Women, Infants, Children] and tacked on their own personal 'compendium' of findings" (9).

The purpose of the WIC program was to improve pregnant women's health and possibly to improve the mental abilities of children by providing a more nutritious early diet. To evaluate the program, the USDA instituted a five-year, $6 million epidemiological study. The sequence of events, which we quote from *SCIENCE*, is a study in the machinations of power-drunk governmental officials.

The original study reported in its summary that the WIC program seemed to be improving fetal health. The technical staff of the Department of Agriculture approved the work and the findings. However, the director of Food and Nutrition and his special assistant—both political appointees—said that the data were too ambiguous to support this conclusion, which they called "an overstatement." They then removed the original summary and inserted new text. This is the power-drunk approach in action.

The next phase involved the closet approach. The head of the study, nutrition expert Dr. David Rush, was never given a chance to comment on the new text of the summary, nor was he asked to review it. The expert advisory committee that was established to oversee the study was also denied the opportunity to review the text of the new summary.

The head of the study, who protested, was told not to testify to the Senate about his complaints. Two years later, using private funds, he published the original work in a peer-reviewed journal. Later, the Department of Agriculture released the original work in unexpurgated form. The two political appointees left government service. Asked by Congress to investigate, the General Accounting Office commented:

> "I have not seen as blatant an example [of tampering] as this in 20 years," said [the GAO director of planning and reporting. . . . [He] said that it was impossible to determine a motive for the USDA's actions, but the result was that "a reader of the compendium would have been misled. . . ." The revised text "did not fairly represent what the study team found," because it was loaded with technical flaws [and it] "did not preserve the original research design." It "provided summary statistics that were not accurate. . . . [and] incorrectly reported conflicts in significance of outcomes" [9].

Others were less timid about attributing motives. Dr. Rush said that "it was clear to him that the former USDA executives wanted to "suppress" the report of WIC's effectiveness. The head of an agency advocating continuation of WIC observed that at the time the study showed positive effects of WIC, the administration had impounded WIC funds and was trying to have them terminated.

Who were the losers? None of the principals seem to have suffered. However, it is now too late to conduct a follow-up on some of the possible positive effects of the program suggested by the study, such as increased skull growth and mental development of children born to WIC mothers. How much satisfaction can these mothers take from knowing that the Department of Agriculture admitted that its political appointee violated the integrity of a major epidemiological study? Is there any reason to believe that it will not happen again?

Some government misuses start modestly enough, but even innocence is no protection when a power-drunk government has a stake in a statistical debate.

Is It Ever Too Late to Get Married?

One of the big news stories of the late 1980s was the "marriage crunch," or "man shortage." The data that caused the considerable press coverage and consequent panic, derived from a report by a Yale sociologist that never-married women who went to college had only a 20% chance of being married at 30, a 5% chance at 35, and a dismaying 1.3% at 40 (10).

At the peak of the feeding frenzy on the news, a research team consisting of the sociologist, a Harvard economist, and a graduate student released a study on the subject. They argued that because of demographic factors, women who went to school before getting married would be at a considerable disadvantage and were, as time went by, finding it necessary to marry increasingly older men. Never mind that at the time a study by the National Center of Health Statistics showed that the age gap was actually *decreasing* (11).

There were a number of flaws in the papers the team produced, but our concern is with the role of the government in the debate. Jeanne Moorman, a demographer in the U.S. Bureau of the Census with a responsibility for tracking marriage and family statistics, carried out her own study using a 1980 sample over two hundred times larger than the 1982 sample used by the research team. The results from this analysis and a review of more recent data contradicted the statistics that apparently showed the existence of a marriage crunch.

Moorman engaged in a dialogue of sorts with the team. She informed them she had put her findings disproving the existence of the marriage crunch into a paper that she would present at the upcoming Population Association of America Conference. The Harvard-Yale team was also to present their paper at the same meeting.

We will not speculate on the reasons for a high-level governmental concern, but:

> Meanwhile at the Census Bureau, Moorman recalls, she was running into interference from Reagan administration officials. The head office handed down a directive, ordering her to quit speaking to the press about the marriage study because such critiques were "too controversial." When a few TV shows actually invited her to tell the other side of the man-shortage story, she had to turn them down. She was told to concentrate instead on a study that the White House wanted—about how poor unwed mothers abuse the welfare system [11, p. 12].

Several months later, at the Population Association of America Conference meeting, both the Harvard-Yale team and Moorman were to present their papers at the same session. The team limited their statements and described their findings as preliminary. Moorman had little to say on the subject:

> Moorman was up next. But thanks to still more interference from her superiors in Washington, there was little she could say. The director of the Census Bureau, looking to avoid further controversy, had ordered her to remove all references to the Harvard-Yale marriage study from her conference speech [11].

Why was it so important to the administration that there be no challenge to the concept of a marriage crunch? Did someone in power want to discourage young women from attending college? Or was it simply that a timid bureaucrat at the Census Bureau really did not want to deal with "controversial" issues?

A Matter of Semantics?

Long, long ago, that is prior to 1971, when releasing new labor statistics, the Commissioner of the Bureau of Labor Statistics (BLS) held press briefings at which professional statisticians interpreted the statistics released by the BLS. Of course, if their explanations were contrary to what the politicians wanted, there was conflict and open discussion in the public arena, but the politicians held the whip hand.

In early 1972, the politicians in power flexed their muscles and discontinued the briefings. According to Lazare Teper,

> The discontinuance of BLS briefings came immediately after two episodes. . . . On February 5, 1971, in the course of the BLS press briefing, Assistant Commissioner Harold Goldstein described a 0.2 percentage point January decline in the unemployment rate as "marginally significant" while simultaneously the Secretary of Labor informed reporters at the White House that the drop had "great significance." At the March 5, 1971 briefing Goldstein viewed another 0.2 percent decline in the unemployment rate, which was accompanied by a drop in employment and hours of work, as being "sort of mixed," while simultaneously the Secretary of Labor described the situation as "heartening" and the Secretary of the Treasury as one that is "rather pleasant." . . . But even suspension of press briefings did not eliminate possible conflict between the professional statisticians and the Administration. In its June 1971 press release, BLS explained that the reported decline in the seasonally adjusted unemployment rate may have been "somewhat overstated because of the seasonal adjustment procedures and because more young workers than usual were in school during the survey week" and that the updating of the seasonal adjustment factors will probably moderate the reported change, as indeed did occur. These valid technical comments apparently produced a reaction within the Administration. President Nixon was apparently furious and his views were conveyed privately to the Department of Labor [12].

Was this no more than a semantic confusion over the difference between the meaning of the word "significance" to a statistician and to a lay person? If Commissioner Goldstein was referring to *statistical* significance, then he meant that, since the reported reduction of 0.2 percent was a sample statistic, it was quite unlikely that the number of U.S. unemployed had dropped. In that case, the Secretary of Labor had little basis for regarding this as a real drop, let alone a "greatly" significant drop.

In the unlikely event that Commissioner Goldstein (a professional statistician) was making a lay judgment of the significance of an estimated reduction of about 10,000 of the nearly 5 million unemployed persons, then many subjective factors enter into the discussion, such as whether you feel this was a "marginally significant" or a greatly significant reduction. The official estimate for 1971 of the number of unemployed persons was 4,993,000; and of employed persons, 86,929,000 (13, Tables 341 and 351).

Whether the basis for these conflicting interpretations was the difference between professional and lay interpretations of the word "significance," or between the Commissioner's and the Administration's subjective points of view, the Administration's actions speak for themselves: The briefings were discontinued.

Shortly after, Harold Goldstein left the Bureau of Labor Statistics, taking early retirement.

Good Enough to Make Nuclear Weapons, But Not Good Enough to Count the Citizenry

As we have mentioned elsewhere, undercounting in the U.S. Census is an ongoing problem. The magnitude of the problem became significant at about the same time that the technical means to resolve the issue became thoroughly established. As discussed by Barbara Bailar, former Associate Director for Statistical Standards and Methodology at the U.S. Bureau of the Census from 1979 to 1987:

> After the 1970 census, the concern about the undercount became more audible, primarily because the state and local governments saw that revenue sharing was tied to census counts. We hear the first rumblings from mayors. . . . It came to light that there was a significant problem of an undercount arising from people living in housing units that have been recorded as vacant. There was another project that showed losses in rural areas. . . . As the Bureau moved toward the 1980 census, no one anticipated lawsuits [14].

But there were lawsuits and they have continued. Their consequences persist:

> The more serious aftermath [of the lawsuits and machinations related to dealing with the undercount] was that the Bureau [of the Census] was seen as severely weakened. It has no power to make its own technical decisions. Major decisions are seen as political and made outside the Bureau [14].

Statistically, the problem is challenging, but not outside the normal range of statistical work. We know now that it is not only an undercount that is the problem; there is *overcounting* in some Census tracts. There are many technical ways to deal with these problems. Of the lot, statistical sampling seems to be the best technical resolution.

The present status (late 1997) is that the Bureau of the Census planned to use sampling to correct the undercount. Is sampling a misbegotten approach? Not if you do it correctly. In 1952, *Railway Age* carried an article reporting on experiments to determine if a sample of waybills could give a reliable estimate of revenue due the Chesapeake and Ohio Railroad for the allocation of charges among several railroads (15). Manufacturers and service providers extensively use sampling to achieve high product and services quality. Without sampling, the production costs of nuclear missiles might well have been prohibitively high. Major businesses and the politicians themselves base multi-million-dollar campaigns on the results of sample surveys. The Census Bureau has its special problems in the field, but no one doubts the technical validity of properly conducted sampling. According to Steven Holmes, writing about two communities, one poor and black, the other well-to-do and white:

> And there is the rub. To Republicans in Congress, sampling has become a dirty word. They say it invites manipulation of figures for partisan purposes and insist that the Constitution requires an actual count of every resident.
>
> What those Republicans rarely add is that many of those undercounted [census] tracts . . . are overwhelmingly Democratic and many tracts that were overcounted . . . are solidly Republican. Republicans fear that adjustments for undercounts and overcounts could cost them Congressional seats [16].

Will the undercounts be corrected in some future census? Currently (1997), some members of Congress want to use the power-drunk approach, exercising the power of legislation. Some members of Congress have threatened to forbid the bureau to supplement counts with estimates based on samples, often using arguments about the technical deficiencies of sampling. But there are some sensible voices heard on the Hill. Here is how Christopher Shays, a Republican member of Congress from Connecticut, sees it:

> Some have raised concerns about the constitutionality of sampling given the requirement for an "actual enumeration." It is likely that any decision on sampling will be the subject of litigation, but I believe that the supports of sampling are on sound legal ground. Stuart Gerson, Assistant Attorney General in the Bush Administration, argues strongly that the framers sought to ensure an accurate Census, which would allow for sampling.
>
> *Unfortunately, some have made the census a partisan issue* [our emphasis]. The census should not be conducted on what party thinks it will win or lose. What matters is that we have an accurate count [17].

Even someone as astute as William Safire of *The New York Times* shows his lack of statistical sophistication when, in an op-ed article (Dec. 7, 1997), he seems to be confusing the difference from sampling of opinions, which involves subjective polling, with sampling of the population for enumeration purposes, which involves counting bodies. Several days later, the Letters to the Editor column set him straight.

If the Undercount Gets Too Messy, the U.S. Could Always Take a Turkish Bath

The U.S. is not the only country that has a potential census undercount problem, but Turkey has for years relied on a unique solution. The Turkish method is quite simple in both concept and execution: on the day of the national census, all Turks stay at home and wait for the census workers to count them. The penalty for not staying at home is Draconian, six months in jail.

The penalty was not really needed for the special census taken on November 30, 1997, since the Turks know that the results of this census will be used to draw electoral districts and apportion federal aid.

> Eager for the greatest amounts of aid, mayors across the country sought to be sure that everyone remained at home to be counted. . . . [the mayor of] Ankara warned last week that if even a single member of any family were found absent today without permission, he would punish the family with a 100 percent surcharge on its natural gas bill and a 600 percent surcharge on its water bill [18].

What do the sequestered Turks do on their national day off? " 'A lot of people are spending the day in bed,' said a police officer. . . . 'Most of these folks work very hard. They can use a day off' [18]." Not everyone is so enthused about this nearly perfect method for getting everyone counted:

> Many census takers and local officials agreed that it is neither logical nor efficient to try to knock on every door in the country on a single day. . . .
> Computer specialists, many of whom viewed today's exercise as a throwback to the Pleistocene epoch, are even more impatient. They say Turkey is lagging far behind in the use of sampling and other demographic tools . . . [18].

We wonder how these computer specialists view the American congressmen trying so hard to prevent the use of sampling in the decennial U.S. Census? Of course, we could always try the Turkish method: shut down the country and send out the bean counters.

SMOKE AND MIRRORS: ECTOPLASTISTICS IN THE SERVICE OF POWER AND POLITICS

Not Accidentally . . .

On November 17, 1982, the then Secretary of Labor, Raymond Donovan, issued a release in which he claimed that the workplace injury rate for the first year of the Reagan Administration (1981) had decreased significantly since the last year of the Carter Administration: "Nearly every major indicator of on-the-job safety and health improved significantly. . . . During the first year of the Reagan Administration workers were less likely to be hurt" (19).

On the same day, the Department of Labor released a statistical report entitled "Occupational Injuries and Illnesses in 1981," prepared by professional statisticians and giving no one any credit for anything. This report simply revealed that the statistical injury rate fell a little more in the last year of the Carter Administration than in the first year of the Reagan Administration. Who should get more credit—Carter, Reagan, or neither? The year-to-year changes in the injury rate are about what one would expect from random fluctuations. What is more likely is that nothing happened between 1980 and 1981 and that neither President deserved credit or blame.

Different Lands, Same Customs

Washington is willing to share its skills in misusing statistics to make government look good with learners from the Third World. The journalist and writer Richard Reeves gives us an example when discussing Pakistani statistics (20, p. 39 ff.). Pakistan's Minister for Planning and Development had worked for the World Bank, an agency of the U.N., for 12 years and had been director of the Policy Planning and Program Review Department. This experience qualified him to design economic plans for the Pakistani government. When Reeves visited him, the minister talked about Pakistan's great economic progress. The minister saw enormous improvement when he compared conditions in 1970 (when he left for Washington) with those in 1982 (when he returned). To support his contentions, he gave Reeves a copy of Pakistan's Sixth Plan.

Reeves found that the statistics shown in the Sixth Plan did indeed show great improvement. And the reason? East Pakistan, now known as Bangladesh, was included in the earlier figures but not in the later ones. This region had been the poorest part of the combined country. When it separated from West Pakistan to become Bangladesh, it took its poverty, malnutrition, and underdevelopment with it. Yes, Pakistan has improved its condition—by surgical removal of the most impoverished part of the country.

Information from the United Nations Annual Yearbook supports Reeves and our analysis of the reason for Pakistan's "improvement." In 1970, income per person in Pakistan was reported to be about $175 (U.S. dollars), and in Bangladesh, about half that amount, $80.* In 1968, when East Pakistan was included with West Pakistan in the national accounts, reported income

---

*Although Bangladesh separated from West Pakistan in 1971, the economists estimated income per person for both West Pakistan and Bangladesh as of 1970. We could not find separate data for East Pakistan (now Bangladesh) prior to 1970.

per person was $120 (21, International Table 1A for 1968 income; 21, Table 1 for 1970 income). Thus, in two years—1968 to 1970—the income per person purportedly increased by 50%. We cannot explain this by any economic activity, including inflation. We can explain such *statistics* in terms of Bangladesh and its poor being included in the earlier computations and then being removed after the political breakaway.

Confucius Says . . .

Since the days of the ancient Chinese philosophers Confucius and Mencius, increases in population have been thought to indicate good government. The ruler who took good care of his subjects thereby increased their numbers, and increases in their numbers was *prima facie* evidence of good and proper government.

From the the beginning of organized government until the 18th century, Chinese governments counted only the people they cared about, those whom they could tax or impress into service for the government.

> In 1712, however, the famous emperor Kang Hsi, of the Tsing Dynasty, detected the gross errors in the population reports. He then decreed that population figures no longer be used as basis for the allotment of the poll and land taxes. This gave local officials greater freedom to boost population growth, especially in the provinces. They could now boast of the apparent material prosperity which was commonly associated with a large and growing population. From this time on, errors of another category appeared because of the tendency deliberately to exaggerate the numbers of people so as to please the reigning emperor [22, p. 1,2].

The purpose of governmental statistics is to please the rulers and keep them in power. In 1977, Leo A. Orleans (23) summarized available information and concluded that there were several forces working against the proper use of statistics in China. He felt that one major force is what might be called "government pride and face." He argued that at that time, the Chinese government was "extremely sensitive to anything that could be interpreted as a failure or even a weakness. Since statistics are the basic measure of success, their publication is closely controlled. Both as an aspect of security and as a manifestation of pride . . ." (23, p. 47).

By 1979, as was indicated by *The New York Times* headline "Press in China admits to lies, boasts and puffery," Chinese newspapers were admitting that many of the success stories they had printed (reporting, for example, huge fruit production increases) were boastful and untruthful, but that they were what the government wanted to hear (24).

Fox Butterfield, who spoke Chinese, interviewed members of the State Statistical Bureau of the People's Republic of China when he was a *New York*

Times reporter in China, and supported this contention. He found the professional statisticians thought that: (a) many factories reported their planned targets rather than their actual production, as plans were often higher than actual production; and (b) agricultural statistics always showed increasing yields per acre as was desired by the government, and this was achieved by reporting less land in use than was actually farmed. The professional statisticians lacked, however, enough personnel to check all reports they received (25).

The catalog is long. Another *New York Times* reporter, C.S. Wren, found similar problems five years later. In 1984, the Chinese press admitted printing false statistics about the Dazhai commune which made it look like one of the most productive communes in the country (26).

A year later, Chinese officials claimed that the governmental policies of economic liberalization attracted $8 billion of foreign capital. But how was "capital" defined? Western embassies that tracked foreign investment in China estimated that the amount of invested foreign capital was less than $1 billion (27). Had trade contracts and other financial arrangements—admittedly of great importance in China's international commercial relationships—accounted for the additional $7 billion? If so, they did not involve actual capital investment by foreigners in our sense of the term. Was this the consequence of legitimate differences between Chinese and Western definitions of capital? Or do they not deserve forgiveness, for they knew what they did?

In the People's Republic of China, the government controls the newspapers; in the United States, the government does not control the newspapers. But it does control the budget of the statistical agencies and sometimes behaves the same way as the Chinese and other governments do in trying to be the power behind the headline. It is the informed reader, like you, who can identify and analyze governmental misuses of statistics and who must try to keep government honest.

15
Afterword

It is wrong always, everywhere and for everyone to believe anything upon insufficient evidence.

—W. K. Clifford

Finding the occasional straw of truth awash in a great ocean of confusion and bamboozle requires intelligence, vigilance, dedication and courage. But if we don't practice these tough habits of thought, we cannot hope to solve the truly serious problems that face us . . . and we risk becoming a nation of suckers, up for grabs by the next charlatan that comes along.

—Carl Sagan

William F. Ogburn, president of the American Statistical Association in 1931, former editor of the *Journal of the American Statistical Association,* and a teacher at the University of Chicago, taught his students to ask three questions to determine the quality of an investigation and whether statistics were being used properly or not:

1. *What are you trying to find out?* What is the problem? Unless the aim of the work or question asked in the report or survey is clearly stated, forget any purported findings.

2. *How do you know it?* Once you have gotten the presumed answer, how do you know it? If statistics have been misused in the process of getting the answer, you do not necessarily have a correct answer and you probably have a wrong answer.

3. *What of it?* Once you have stated the problem to be solved or questions to be answered and obtained a solution or an answer, the question of importance and relevance must be answered. Was the problem important enough to be worth the effort?

In general, we have not discussed questions 1 and 3 in our book. These two questions are judgmental and often can be answered only in a particular context. Different people are interested in different subjects at different times, and both intent and importance are usually subjective. As long as the writer has clearly stated the problem which he or she seeks to answer, we have no complaints. How many "dope addicts" are there? How many people keep scorpions as pets? What is the balance of trade? How many high school dropouts are unemployed? Pick the question or problem which is relevant to you. Use the results as you see fit. We are only concerned that, if you or someone else uses statistics to get answers, the appropriate statistics be used correctly so that as few people as possible are fooled.

Be vigilant and scrutinize!

References

CHAPTER 1: INTRODUCTION

1. I. Bernard Cohen, Florence Nightingale, *Scientific American, 250*(3):128 (1984).
2. H. Friedman and Linda W. Friedman, A new approach to teaching statistics: learning from misuses, *New York Statistician, 31*(4–5):1 (1980).
3. Stephen Jay Gould, Singapore's patrimony (and matrimony), *Natural History, 93*(5):22 (1984).
4. Herbert F. Spirer and A. J. Jaffe, *New York Statistician, 32*(3):1 (1981).
5. Stephen Jay Gould, Morton's ranking of races by cranial capacity, *Science, 200*(4341):503 (1978).
6. Leonard Silk, Economic scene. Call to hire 'One Person,' *New York Times,* December 29, 1982.
7. Louis Pollack and H. Weiss, Communication satellites: countdown for Intelsat VI, *Science, 223*(4636):553 (1984).
8. Letters to the editor, *Science, 224*(4648):446 (1984).

CHAPTER 2: CATEGORIES OF MISUSE

1. Phony formulas, *Wall Street Journal,* April 2, 1979.
2. Zero Population Growth, Washington, D.C., National Referendum, undated (received in 1996).
3. E. J. Dionne Jr., Abortion poll: not clearcut, *New York Times,* Aug. 18, 1980.
4. *United States Code 1994,* v. 23, Title 42—The Public Health and Welfare, U.S. Government Printing Office, Washington, §11302. General definition of homeless individual.
5. Rep. David Dreier, Letter to the editor, *Wall Street Journal,* May 10, 1990.
6. Loy I. Julius, Richard W. Hungerford, William J. Nelson, Theodore McKercher, and Robert W. Zellhoefer, Prevention of dry socket with local application of TerraCortril in Gelfoam, *Journal of American Oral and Maxillofacial Surgeons, 40*:285 (1982).
7. Herbert F. Spirer, S. Rappaport, and A. J. Jaffe, Misuses of statistics: the jaws of man I, *New York Statistician, 34*(5):3 (1983).

8. Alice Felt Tyler, Freedom's Ferment: Phases of American Social History from the Colonial Period to the Outbreak of Civil War, Harper Torchbooks, New York, 1944, p. 229.
9. Minimum-competency tests having minimum effects, *New York Times,* April 22, 1979.
10. John Walsh, Test scores—Are they a distorted proxy for achievement?, *Science, 237*(4819):1100 (1987), p. 7.
11. Malcolm W. Browne, When numbers just don't add up, *New York Times,* April 22, 1980.
12. Jerry E. Bishop, Surgeons find heart repair pays for itself, *Wall Street Journal,* Aug. 29, 1980.
13. Glenn Collins, Migration trend of the aging, *New York Times,* Oct. 2, 1984.
14. Karen W. Arenson, Martin Feldstein's computer error, *New York Times,* Oct. 5, 1980.
15. Sam Rosensohn, Happiness is . . . being able to say 'no' to sex, *New York Post,* Sept. 16, 1980.
16. Still sexy after all those years, *Modern Maturity*, February-March 1988, p. 11.
17. Lisa Belkin, The 900 number as audience pollster, *New York Times*, Jul 6, 1987.
18. Joyce Purnick, Convictions of 3.8% found in updated day-care check, *New York Times,* Feb. 18, 1985.
19. Advertisement, *New York Times Magazine,* July 15, 1979.
20. Shanna H. Swan and Willard I. Brown, Vasectomy and cancer of the cervix, *New England Journal of Medicine, 301:*46 (1979).
21. Phillip Shabecoff, Jobless rate in U.S. up slightly from July, to 5.7% from 5.6%, *New York Times,* Aug. 4, 1979.
22. Gina Kolata, Computing in the language of science, *Science, 224*(4645):150 (1984).

CHAPTER 3: KNOW THE SUBJECT MATTER

1. U.S. Bureau of the Census, *1970 Census,* Vol. PC(2)-1F, p. X.
2. Peter Hubbard, *Science, 217*(4563):919 (1982).
3. *New York Times*, Oct. 30, 1979.
4. U.S. Bureau of the Census, *1970 Census*, Vol. PC(2)-7A.
5. U.S. Office of Education, *Earned Degrees.* (See annual issues.)
6. Barbara Curran, ed., *Women in the Law: A Look at the Numbers*, American Bar Association, Commission on Women in the Profession, Washington, 1995.
7. *Judicial Selection Project: Annual Report 1996*, Alliance for Justice, Washington, 1996.
8. Robert Lewis, A pension crisis that wasn't: Aging committee allegedly 'misinterpreted' data, *AARP Bulletin*, Sept. 1997, p. 14.
9. U.S. Bureau of the Census, Nation to reach zero population growth by 2050, Release, Nov. 9, 1982.
10. Bernard Brown, *Science*, March 21, 1997, p. 1425.

11. U.S. Bureau of the Census, *Statistical Abstract of the United States: 1996*, Tables 14, 15, 17 (middle series).

CHAPTER 4: DEFINITIONS

1. U.S. Bureau of the Census, *Statistical Abstract of the United States: 1996*.
2. U.S. Bureau of the Census, *1980 Census of Population, Characteristics of the Population, General Social and Economic Characteristics U.S. Summary*, Vol. 1, Chapter C, Part 1, PC 80-1-C1, 1983.
3. Constitution of the United States, Article I, Section II, Part 3.
4. No Multi-Racial on Census [online]. National Public Radio, All Things Considered. July 8, 1997. Available from World Wide Web: <URL:http://www.npr.org/news/national/970708/multiracial.html>.
5. Producers' prices show slight drop, first in 4½ Years, *New York Times,* Oct. 4, 1980.
6. Keith Bradsher, Average car price $20,000? Not quite, it seems, *New York Times*, Jan. 18, 1996.
7. Constance F. Citro and Robert T. Michael (eds.), *Measuring Poverty: A New Approach*, Washington: National Academy Press, 1995.
8. David Cay Johnston, In policy shift, tax auditor's scrutiny falls unexpectedly, *New York Times*, April 13, 1997, p. 26)
9. United Nations, *Statistical Year Book,* New York, 1981.
10. Alan Finder, Candidates For Senate Use Truth, Of a Sort, *New York Times,* Oct. 7, 1996.
11. Keith Harriston and Avis Thomas-Lester, Murder Capital Reputation Derided By New D.C. Chief, *Washington Post*, Jan. 1, 1993. *New York Times,* Jan. 13, 1980.
12. Sandra R. Gregg, Mayor announces 30% drop in TB rate, *Washington Post,* April 14, 1982.
13. U.S. Department of Health and Human Services, *Monthly Vital Statistics Report*, v. 43, no. 9, Supplement, March 22, 1995.
14. Johnathan Lucas, Lab error blamed in faulty mercury report, *The Advocate*, Feb. 15, 1997.
15. Leon J. Kamin, BOOK REVIEW: Behind the curve, *Scientific American*, 272(2):100 (1995).
16. Douglas A. Samuelson, What is intelligence? A re-examination, presented at the 1976 Annual Meetings, American Statistical Association.

CHAPTER 5: THE QUALITY OF BASIC DATA

1. Max Singer, The vitality of mythical numbers, *The Public Interest, 23*:3 (1971).
2. New satellite data hint that universe may be bigger than thought, *New York Times*, Feb. 15, 1997.

3. Alan Finder, City's survey of cleanliness of its streets questioned, *New York Times,* May 14, 1997.

4. John Head, 20 million illegal aliens get reprieve, *Denver Post,* Sept. 7, 1980.

5. Jeffrey S. Passel, Undocumented Immigrants: How Many?, 1985 Annual Meetings of the American Statistical Association, Las Vegas, 1985.

6. *Philadelphia Bulletin,* Oct. 21, 1979.

7. Melvin A. Benarde, Letter to the editor: food additives, *Science, 206*(4424):206 (1979).

8. Kathleen Stein, Dr. C's vitamin elixirs, *OMNI* (issue unknown), p. 69 (1982).

9. U.S. Public Health Service, *U.S. Life Tables, 1949–51, Vital Statistics Special Reports, 41*(1):8 (1954).

10. U.S. Bureau of the Census, *Statistical Abstract of the United States: 1984.*

11. U.S. Federal Bureau of Investigation, *Uniform Crime Reporting Handbook,* 1966.

12. Lawrence Sherman and Barry Glick, The quality of police arrest statistics, *PF Reports,* Aug. 1984, p. 1.

13. F.B.I. Crime survey drops Philadelphia, *New York Times,* Oct. 20, 1997.

14. Soybeans soar as U.S. agency says error overstated supply, *New York Times,* April 2, 1997.

15. Can't Brady's bunch count?, *New York Times,* March 1, 1992.

16. Karen Arenson, CUNY trustees question the value of remedial program in math and writing, *New York Times,* March 18, 1997.

17. Paul R. Fish, Consistency in archeological measurement and classification: a pilot study, *American Antiquity, 43*:86 (1978).

18. Sharon L. Bass, Despite law, police fail to report bias crimes, *New York Times,* April 23, 1989.

19. A. J. Jaffe, *The First Immigrants from Asia: A Population History of the North American Indians,* Plenum, New York, 1992.

20. U.S. Bureau of the Census, *Statistical Abstract of the United States: 1996.*

21. U.S. Bureau of the Census, *Statistical Abstract of the United States: 1916.*

22. A. M. Gibson, The American Indian: Prehistory to the present, Norman: University of Oklahoma Press, 1980.

23. Tony Hillerman, Who has sovereignty over Mother Earth?, *New York Times,* Sept. 18, 1997, p. A35.

24. Bureau of Applied Social Research of Columbia University, *Puerto Rican Population of New York City* (A. J. Jaffe, ed.), New York, 1954.

25. Survey reports fertility levels plummet in developing nations, *New York Times,* Aug. 10, 1979.

26. Joseph A. Cavanaugh, Is fertility declining in less developed countries? An evaluation analysis of data sources and population programme assistance, *Population Studies, 33*(2):283 (1979).

27. Alfred L. Malabre, Thanks to off-the-books income, consumers save more than meets the eye, economists say, *Wall Street Journal,* June 11, 1982.

28. Denise B. Kandel, ed., *Longitudinal Research on Drug Use,* Wiley, New York, 1978.

29. John A. O'Donnell, Variables affecting drug use (book review), *Science, 203*(4382):739 (1979).

30. Denise Kandel, Stages in adolescent involvement in drug use, *Science, 190*:912 (1975).
31. Joseph A. Raelin, *Building a Career: The Effect of Initial Job Experiences and Related Work Attitudes on Later Employment*, Upjohn Institute for Employment Research, Kalamazoo, Mich., 1980.
32. Louis Uchitelle, The sudden shifts of forecasters, *New York Times*, Sept. 1, 1989.
33. U.S. Bureau of the Census, *Statistical Abstract of the United States: 1990.*
34. Anonymous reviewer, private communication, Sept. 1985.
35. Sey Chassler, The Redbook report on sexual relationships, *Redbook Magazine*, Oct. 1980.
36. Roger S. Bagnall, For young classicists, a silver lining, *New York Times*, Sept. 18, 1985.

CHAPTER 6: GRAPHICS AND PRESENTATION

1. William S. Cleveland, *The Elements of Graphing Data*, Hobart Press, Summit, N.J., 1994.
2. Edward R. Tufte, *The Visual Display of Quantitative Information,* Graphics Press, Cheshire, Conn., 1983*; Envisioning Information,* Graphics Press, Cheshire, Conn., 1990; *Visual Explanations, Graphics Press*, Cheshire, Conn., 1997.
3. Howard Wainer, *Visual Revelations: Graphical Tales of Fate and Deception from Napoleon Bonaparte to Ross Perot.* New York: Copernicus Books, 1997, Chapters 1 and 2.
4. Martin Mayer, Tremor in Orange County, *Barron's,* Jan. 16, 1978.
5. U.S. Department of Labor, Bureau of Labor Statistics, *Consumer Price Index for Urban Consumers (CPI-U)*, U.S. city average, All items, Series A, 1979.
6. Where two aggressive companies plot growth, *Business Week,* June 16, 1980.
7. Howard Wainer, Visual revelations: Measuring graphicy, *Chance V 3*(4), 52 (1990).
8. Energy Information Administration/Annual Energy Review 1995, National Energy Information Center, p. 9.
9. Frederick C. Klein, Americans hold increasing amounts in cash despite inflation and many other drawbacks, *Wall Street Journal,* July 5, 1979.
10. *A.F.L.-C.I.O. News,* Feb. 3, 1979.
11. Steven J. Marcus, Solar-age windows, *New York Times*, April 21, 1983.
12. U.S. Bureau of the Census, *Statistical Abstract of the United States: 1984.*
13. James Black, Misuse of statistics 27—Representing a loss as a gain, *New York Statistician, 35*(4):3 (1984).

CHAPTER 7: METHODOLOGY

1. U.S. Bureau of the Census, *Statistical Abstract of the United States: 1996.*
2. Martin Gottlieb and Kurt Eichenwald, A hospital chain's brass knuckles, and the backlash, *New York Times*, May 11, 1997.

3. Karen W. Arenson, Economics: Martin Feldstein's computer error, *New York Times,* Oct. 5, 1980.

4. Leonard Silk, Economic scene: Social Security impact a puzzle, *New York Times,* Dec. 17, 1982.

5. Dean R. Leimer and Selig D. Lesnoy, What they found . . . , *New York Times,* Oct. 5, 1980.

6. Martin S. Feldstein, . . . And his defense, *New York Times,* Oct. 5, 1980.

7. Daniel Patrick Moynihan, North Dakota, math country, *New York Times*, Feb. 3, 1992.

8. *NAEP 1996 Mathematics Report Card*, National Assessment of Educational Progress, National Center for Education Statistics, Washington (1996), p. 30.

9. Frederick Mosteller and John W. Tukey, *Data Analysis and Regression,* Addison-Wesley, Reading, Mass., 1977.

10. News Releases, Filings by Chapter of the Bankruptcy Code—by Quarter, various years, Administrative Office of the U.S. Courts, Bankruptcy Division, Washington.

11. Federal Reserve statistical releases for various years, Board of Governors of the Federal Reserve System, Washington.

12. Barbara Pitcher, *Summary Report of Validity Studies Carried Out by ETS for Graduate Schools of Business 1954–1970,* Educational Testing Service, Princeton, N.J., 1971.

13. Sandra Rosenhouse-Persson and George Sabagh, Attitudes toward abortion among Catholic Mexican-American women: the effects of religiosity and education, *Demography*, 20(1):87 (1983).

14. Growth hormone fails to reverse effects of aging, researchers say, *New York Times*, April 15, 1997.

15. J. Neyman and E. S. Pearson, *Joint Statistical Papers,* University of California Press, Berkeley, 1967.

CHAPTER 8: FAULTY INTERPRETATION

1. Fujiya Hotel, Ltd., *We Japanese, Being Descriptions of Many of the Customs, Manners, Ceremonies, Festivals, Arts and Crafts of the Japanese,* Miyanoshita, Japan, 1950.

2. Constance F. Citro and Robert T. Michael (eds.), Measuring Poverty: A New Approach, National Academy Press: Washington, 1995, p. 3.

3. John D. Durand, Development of the labor force concept 1930–1940, *Labor Force Definition and Measurement,* Social Science Research Council, New York, 1947.

4. A. J. Jaffe and Charles D. Stewart, *Manpower Resources and Utilization,* Wiley, New York, 1951.

5. U.S. Department of Labor, *Employment and Earnings,* 1985.

6. U.S. Department of Labor, *Employment and Earnings,* 1983.

7. M. T. Kaufman, No longer a charity case, India fills its own granaries, *New York Times,* Aug. 10, 1980.

8. *World Development Report,* Oxford University Press, New York, 1983, Table 24.
9. U.S. Bureau of the Census, *Statistical Abstract of the United States: 1963.*
10. United Nations Food and Agriculture Organization, 1984 *State of Food and Agriculture,* New York, 1984, Annex Tables 2 (food production) and 3 (total agricultural production).
11. Raymond Bonner, Gains from El Salvador land distribution disputed, *New York Times,* Apr. 19, 1982.
12. Edwin B. Williams (ed.), *The Scribner—Bantam English Dictionary,* Charles Scribner's Sons, New York, 1977.
13. Mark L. Trencher, Misuse of statistics 21, *New York Statistician, 34*(1):5 (1982).
14. AFL–CIO Public Employee Department, *REVENEWS,* Feb. 1982.
15. Franklin L. Leonard, Misuse of statistics XIV, *New York Statistician, 32*(4):2 (1981).
16. Stephen R. Kellert and Alan R. Felthaus, Childhood cruelty toward animals among criminals and noncriminals, manuscript submitted to *Archives of General Psychiatry,* November 1983. Private communication.
17. W. Edwards Deming, Letter to Prof. William Cochran, Harvard University, Feb. 19, 1972.
18. Tom R. Houston, Why models go wrong, *BYTE,* Oct. 1985.
19. Andrew F. Siegel, *Practical Business Statistics,* ed. 2, Irwin, Burr Ridge, IL, 1994.
20. David S. Moore and George P. McCabe, *Introduction to the Practice of Statistics,* W. H. Freeman and Company, New York, 1989.

CHAPTER 9: SURVEYS AND POLLS, PART I

1. U.S. Bureau of the Census, *Statistical Abstract of the United States: 1996.*
2. The man who knows how we think, *Modern Maturity,* April-May, 1974.
3. Archdiocese of New York, Office of Pastoral Research, *Hispanics in New York: Religious, Cultural and Social Experiences,* Vol. 1, New York, 1982.
4. Norman Bowers and Francis W. Horvath, Keeping time: An analysis of errors in the measurement of unemployment duration, *Journal of Business and Economic Statistics,* 2:2 (1984).
5. Federation Employment and Guidance Service of New York, *Survey of Employers' Practices and Policies in the Hiring of Physically Impaired Workers,* A. J. Jaffe (dir.), New York, 1959.
6. Peter S. Temes, Soviet anti-Semitism survey shows bias, *New York Times,* April 16, 1990.
7. Congressman Christopher Shays, Fourth District, Connecticut, 1995 Legislative Questionnaire, 1995.
8. Howard Schuman and Jacqueline Scott, Problems in the use of survey questions to measure public opinion, *Science, 236*:957 (1987).
9. Sam Howe Verhovek, Houston to vote on repeal of affirmative action, *New York Times,* Nov. 2, 1997.

10. Frederick Mosteller, Herbert Hyman, Phillip J. McCarthy, Eli S. Marks, and David Truman, The pre-election polls of 1948, *Science Research Council Bulletin, 60,* New York, 1949.
11. Caroline Bird, The jobs you dream about, *Modern Maturity,* Feb.-Mar. 1988.
12. J. Hachigian, Misuse of statistics XI, Carter-Reagan debate and the ABC telephone poll, *New York Statistician, 32*(2):1 (1980).
13. *Congressman Hollenbeck's Washington Report,* June, 1980, p. 4.
14. A. J. Jaffe, Not everyone has a telephone at home, *New York Statistician, 35*(5):2 (1984).
15. Roger L. Jenkins, Richard C. Reizenstein, and F. G. Rodgers, Report cards on the MBA, *Harvard Business Review, 62*(5):20 (1984).
16. U.S. Bureau of the Census, *Employment and Earnings,* 1983.
18. Century Opinion Polls, Inc., Report of Registered Democrats Regarding the 1980 Presidential Primary, *New York Poll,* May 25, 1979.
19. Research and Forecasts, Inc., *Aging in America: Trials and Triumphs,* New York, 1980.
20. Marian Burros, Major U.S. Survey on food use and pesticides is drawing fire, *New York Times,* Sept. 11, 1991.

CHAPTER 10: SURVEYS AND POLLS, PART II

1. Bill Carter, Does more time on line mean reduced TV time?, *New York Times,* Jan. 31, 1997.
2. U.S. Bureau of the Census, Money Income in the United States, Series P60-193, 1995.
3. Jobs: unemployment is down, but not in Detroit, *New York Times,* Dec. 9, 1979.
4. U.S. Bureau of Labor Statistics, *Geographic Profile of Employment and Unemployment: States, 1978; Metropolitan Areas, 1977–78,* 1978.
5. Have we lost confidence in our healthcare system? 80% say yes, *Physicians Financial News, 15*(3):8 (1997).
6. Robert B. Gunnison, Wilson's image takes clobbering in new state poll, San Francisco Chronicle, Feb. 20, 1997.
7. Philip Converse and Michael Traugott, Assessing the accuracy of polls and surveys, *SCIENCE, 234*:1097 (1986).
8. Michael Kagay and Janet Elder, Numbers are no problem for pollsters, words are, *New York Times,* Aug. 9, 1992.
9. John R. Lott Jr., Viewpoint, The (Stamford, CT) Advocate, July 7, 1997.
10. M. Kristen Rand, Questionable study financed by the gun industry, The (Stamford, CT) Advocate, July 11, 1997.
11. U.S. Bureau of Labor Statistics, *Employment and Earnings,* 1983.
12. Daniel Yankelovich, The polls don't mean a thing, *New York Times,* Oct. 7, 1979.
13. Janet Elder, Poll finds women are the health-savvier sex, and the warier, *New York Times,* June 22, 1997.

14. Barbara A. Bailar and C. Michael Lanphier, *Development of Survey Methods to Assess Survey Practices,* American Statistical Association, Washington, D.C., 1978.
15. A. J. Jaffe, Walter Adams, and Sandra G. Meyers, *Negro Higher Education in the 60s,* Praeger, New York, 1968.

CHAPTER 11: THE LAW OF PARSIMONY

1. John Kenneth Galbraith, in an article in the *Atlantic Monthly* as quoted in the *Wall Street Journal,* issue unknown.
2. Edi Karni and Barbara K. Shapiro, Tales of horror from ivory towers, *Journal of Political Economy,* Feb. 1980.
3. American Anthropological Association, Report of the ad hoc committee to implement the 1972 resolution on fair employment practices in employment of women, *Anthropology Newsletter, 20*:7 (1979).
4. A. J. Jaffe, Jaffe disagrees with committee's procedures, *Anthropology Newsletter, 20*(7):4 (1979).
5. David H. Thomas, The awful truth about statistics in archaeology, *American Antiquity, 43*(2):231 (1978).
6. Joel Gunn, *American Antiquity, 40*(3):21 (1975).
7. W. Brass and A. J. Coale, Methods of analysis and estimation, *The Demography of Tropical Africa,* W. Brass (ed.), Princeton University Press, Princeton, N.J., 1968.
8. W. Brass, indirect methods of estimating mortality illustrated by application to Middle East and North Africa data, *The Population Framework, Demographic Analysis, Population and Development,* UNECWA, Beirut, Lebanon, 1978.
9. W. Brass, *Methods for Estimating Fertility and Mortality from Limited and Defective Data,* Laboratory for Population Studies, University of North Carolina, Chapel Hill, NC, 1975.
10. Michel Garenne, Problems in applying the Brass method in tropical Africa: A case study in rural Senegal, *Genus,* Consiglio Nazionale delle Richerche (ed.), *XXXVIII*(1–2), Comitato Italiano per lo Studio dei Problemi della Popolazione, Rome, 1982, p. 119.
11. Wassily Leontief, Academic economics, Letter to the Editor, *Science, 217*(4555):104 (1982).
12. Richard A. Staley, Nonquantification in economics, Letter to the Editor, *Science, 217*(4566):1204 (1982).
13. Gerald A. Bodner, Point of view: should statistics alone be allowed to prove discrimination? *Chronicle of Higher Education,* June 22, 1983.
14. Wesley A. Fisher, Ethnic consciousness and intermarriage: Correlates of endogamy among the major Soviet nationalities, *Soviet Studies, XXIX*(3):395 (1976).
15. Brian D. Silver, Ethnic intermarriage and ethnic consciousness among Soviet nationalities, *Soviet Studies, XXX*(1):107 (1977).

16. Raymond P. Mayer and Robert A. Stowe, Would you believe 99.9969% explained?, *Industrial and Engineering Chemistry, 61*(5):42 (1969).
17. Thomas Leigh, David MacKay, and Joyhn Summers, Reliability and validity of conjoint analysis and self-explicated weights: A comparison, in Joseph Hair, Rolph Anderson, and Ronal Tatham, *Multivariate Data Analysis with Readings*, ed. 2, Macmillan Publishing Company, New York (1987).
18. Douglas E. Samuelson, E-mail communication, Aug. 23, 1997.
19. O. Felix Offodile, Kingsley O. Ugwu, and Leslie Hayduk, Analysis of the causal structures linking process variables to robot repeatability and accuracy, *Technometrics, 35*, (4) (1993).
20. Philip Holmes-Smith, Introduction to structural equation modeling using LISREL and AMOS, Department of Education (Victoria), National University of Australia [cited Aug. 8, 1997]. Available from the World Wide Web: <URL:http://www.ssda.anu.edu.au/acspri/courses/summer/sp97/LISREL.html>.
21. Anon., Personal communication.

CHAPTER 12: THINKING: LACK OF FORETHOUGHT, LACK OF AFTERTHOUGHT

1. David F. Kerridge, Statistical thinking, Deming NYU Seminar for Statisticians, New York, 1991.
2. Julie Miller, High costs are blamed for the loss of a mill, *New York Times*, Connecticut Section, Dec. 29, 1996.
3. Correction, *New York Times*, Connecticut Section, Jan. 12, 1997.
4. James George Frazer, *The Golden Bough*, Macmillan Company, New York (1945).
5. Robyn Dawes, *Rational Choice in an Uncertain World*, Harcourt Brace Jovanovich, New York (1988).
6. W. C. Thompson and E.L. Schumann, Interpretation of statistical evidence in criminal trials, *Law and Human Behavior, 11*:167 (1987).
7. Bernard Robertson and G. A. Vigneaux, *Interpreting Evidence: Evaluating Forensic Science in the Courtroom*, Wiley, New York (1995).
8. Bumper sticker, Donnely/Colt, Hampton, CT, Jan. 9, 1997.
9. U.S. Bureau of the Census, *Statistical Abstract of the United States: 1996.*
10. Richard Ellis, Death in Athens, Letter to the Editor, *Science, 273*:1037 (1996).
11. J. B. Bury, S. A. Cook, F. E. Adcock, eds. *The Cambridge Ancient History*, Cambridge University Press, Cambridge, UK, 1935.
12. Richard W. Stevenson, U.S. to revise its estimate of layoffs, *New York Times*, Oct. 16, 1996.
13. Anon., *Ayrshire Leader*, Nov. 4, 1993. Supplied by David Finney.
14. Edward MacNeal, The Mathsemantic Monitor, PLAYING THE PERCENTAGES: Sex in Connecticut, *ETC: A Review of General Semantics et cetera, 3*:3(275–284).
15. Virginia Morell, The origin of dogs: Running with the wolves, *Science, 276*(5319):1647–1648 (1997).

16. Robert Shaffer, Where all the children are above average, Letter to the Editor, *New York Times*, June 14, 1997.
17. Peter Applebome, Proposal for school testing is facing wide opposition, *New York Times*, Aug. 3, 1997.
18. John Jacob Cannell, Nationally normed elementary achievement tests in America's public schools: How all fifty states are above the national average, Friends for Education, Daniels WV (1987).
19. Jennifer Egan, The thin red line, *New York Times Magazine*, July 27, 1997.
20. As quoted in Daniel Patrick Moynihan, The big lie of 1996, *Washington Post*, Jan. 28, 1997.
21. Richard Nisbett, Georffrey Fong, Darrin Lehman, Patricia Cheng, Teaching reasoning, *SCIENCE*, *238*:625–631 (1987).

CHAPTER 13: ECTOPLASTISTICS

1. R. J. Rummel, The Hell state: Cambodia under Khmer Rouge, draft chapter, 1991.
2. Rhoda Howard, *Human Rights and the Search for Community*, Westview Press, Boulder, 1995.
3. U.S. Bureau of the Census, *Statistical Abstract of the United States: 1996*.
4. Janus Adams, Who'll pay the price in the Simpson case?, The (Stamford, CT) Advocate, Jan. 12, 1997.
5. A quiz on women's health, The (Stamford, CT) Advocate, March 25, 1997.
6. CCADV, *Connecticut Coalition Against Domestic Violence*, 135 Broad St., Hartford, CT 06105.
7. Lenore Weitzman, *The divorce revolution: The unexpected social and economic consequences for women and children in America*, The Free Press, New York (1985).
8. Nancy Rubin, Middle-aged, single, looking for a life, *New York Times*, Connecticut section, Dec. 7 (1997).
9. Richard Peterson, A re-evaluation of the economic consequences of divorce, American Sociological Review, *61*:528–536 (1996).
10. Leonore J. Weitzman, The economic consequences of divorce are still unequal: Comment on Peterson, *American Sociological Review*, 61:537–538 (1997).
11. Richard Peterson, Statistical errors, faulty conclusions, *American Sociological Review*, 61:539–540 (1996).
12. Verlyn Klinkenborg, Awakening to sleep, *New York Times*, Jan. 5, 1997.
13. 80,000 lobbyists? Probably not, but maybe . . . , *New York Times,* May 20, 1993.
14. Max Singer, The vitality of mythical numbers, *The Public Interest*, *23*:3 (1971).
15. Marc Galantner, Reading the landscape of disputes, *U.C.L.A. Law Review*, *31*:4 (1983).
16. Paul Brodeur, Annals of law, the asbestos industry on trial, Parts I to IV, *The New Yorker,* June 10, 17, 24, and July 1, 1985.

17. James Bennet, On eve of African relief talks, aid donors argue over numbers, *New York Times*, Nov. 22, 1996.
18. Nicholson Baker, The author vs. the library, *The New Yorker*, Oct. 14, 1996.

CHAPTER 14: BIG BROTHER/SISTER

1. Garcilaso de la Vega, El Inca, *Royal Commentaries of the Incas, Part One,* Harold V. Livermore (transl.), University of Texas Press, Austin, 1966.
2. Nicholas F. Jones, *Ancient Greece: State and Society*, Prentice Hall, Upper Saddle River, NJ, 1997, pp. 24–25.
3. C. Taeuber, "Census." *International Encyclopedia of the Social Sciences,* Vol. 2, Crowell, Collier, and Macmillan, New York, 1968.
4. Martin Tolchin, Pick a number, any politically powerful number, *New York Times,* June 5, 1984.
5. *SCOPE,* Creative Journals, Ltd., London, October 1944.
6. M. C. Buer, *Health, Wealth, and Population in the Early Days of the Industrial Revolution,* George Routledge & Sons, London, 1926.
7. Clifford D. May, An ancient disease advances, *New York Times, Aug.* 18, 1985.
8. Fred W. Grupp and Ellen Jones, Infant mortality trends in the Soviet Union, unpublished manuscript, 1983.
9. Eliot Marshall, USDA admits "mistake" in doctoring study, *SCIENCE, 247*(4942TK):522 (1992).
10. Lisa Petersen, They're falling in love again, say marriage counselors, The (Stamford, CT) Advocate, Feb. 14, 1986, as quoted in Susan Faludi, *Backlash: The Undeclared War Against American Women*, Crown Publishers, New York, 1991, p. 9.
11. Susan Faludi, *Backlash: The Undeclared War Against American Women*, Crown Publishers, New York, 1991, p. 10.
12. Lazare Teper, Politicization and statistics, *New York Statistician, 25*(4):1 (1974).
13. U.S. Bureau of the Census, *Statistical Abstract of the United States: 1972.*
14. Barbara Bailar, Rumination on the Census, *AMSTAT NEWS, 245*:1, 3 (1997).
15. Can scientific sampling techniques be used in railroad accounting?, *Railway Age*, June 9, 1952, pp. 61–64.
16. Steven Holmes, 2 communities illustrate debate over Census, *New York Times*, Aug. 30, 1997.
17. Christopher Shays, Letter to Louise Spirer, July 16, 1997.
18. Stephen Kinzer, For one day, every Turk hides in plain sight, *New York Times*, Dec. 1, 1997.
19. A. J. Jaffe, Politicians at play, *New York Statistician, 34*(4):1 (1983).
20. Richard Reeves, Journey to Pakistan, *The New Yorker,* Oct. 1, 1984.
21. United Nations, *United Nations Year Book of National Accounts,* Vol. III, United Nations, New York, 1972.
22. Chen Ta, *Population in Modern China,* University of Chicago Press, Chicago, 1946.

23. Leo A. Orleans, Chinese statistics: the impossible dream, *American Statistician, 28*(2):47 (1974).
24. Press in China admits to lies, boasts and puffery, *New York Times,* Aug. 29, 1979.
25. Fox Butterfield, *China,* Bantam Books, New York, 1983.
26. Christopher S. Wren, Chinese press tries to mend soiled image, *New York Times,* May 20, 1984.
27. John F. Burns, A new team checks in at the Great Wall Hotel, *New York Times,* March 24, 1985.

Index